HOW TO KILL AN ASTEROID

ALSO BY ROBIN GEORGE ANDREWS

Super Volcanoes: What They Reveal about Earth and the Worlds Beyond

HOW TO KILL AN ASTEROID

THE REAL SCIENCE OF PLANETARY DEFENSE

Robin George Andrews

W. W. NORTON & COMPANY

Independent Publishers Since 1923

Printed in the United States of America
First Edition

For information about special discounts for bulk purchases, please contact W. W. Norton Special Sales at specialsales@wwnorton.com or 800-233-4830

Manufacturing by Lakeside Book Company
Book design by Chris Welch
Production manager: Louise Mattarelliano

ISBN: 978-1-324-05019-3

W. W. Norton & Company, Inc.
500 Fifth Avenue, New York, N.Y. 10110
www.wwnorton.com

W. W. Norton & Company Ltd.
15 Carlisle Street, London W1D 3BS

10 9 8 7 6 5 4 3 2 1

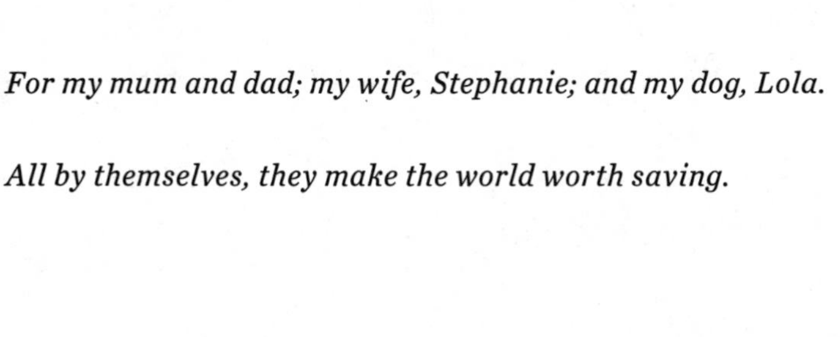

For my mum and dad; my wife, Stephanie; and my dog, Lola.

All by themselves, they make the world worth saving.

CONTENTS

PROLOGUE

Apocalypse Please

It's February 19, 2046. It doesn't have to be. We don't know when this will happen or exactly where on Earth it may strike. But it could very well be February 19, 2046. We're in Seattle. Why not? Bit rainy, but it's a perfectly charming city. And it's a Monday. Apart from a few sociopaths, everybody hates Mondays—and this Monday will be especially awful, because about 10 million years ago, two elephantine asteroids far from our planet's doorstep crashed into each other and broke apart.

Over the epochs, one of those small shards—one just 460 feet across, not much longer than a standard American football field—has been circumnavigating the Sun, crossing Earth's orbital racetrack over and over again, entirely harmlessly. But as myriad millennia vanished into the past, that relatively diminutive asteroid and our blue-green marble have inched closer and closer together. Eventually, despite their fundamentally clumsy and cantankerous natures, humans evolved, created civilizations, and stopped killing each other long enough to build increasingly powerful telescopes and point them at the deep and beautiful dark above. They found plenty of colossal asteroids, and many smaller fragments, too. But some were too small, too fast, too elusive to find. And by the time February 19, 2046, came around, that 460-foot rocky voyager managed what so many of its friends had failed to do: it found us.

By 2046, about 1.6 million people live in this picturesque corner of the Pacific Northwest. For most of them, it's an ordinary Monday morning. Someone's accidentally locked themself out of their house. In its attempt to chase a squirrel, a normally acrobatic cat tumbles off a fence and lands in a bush, startling the woman sitting on the bench nearby. Three friends are chatting over coffee. Someone drives past in an electric car blasting Frank Sinatra's *My Way* out of their rolled-down windows, because this is the day they tell their awful boss they're quitting. And it's sunny: improbably, there's not a cloud in the sky.

That makes it easier to see a bright streak of light leaving a graffiti trail of ashen dust across the azure. Anyone that spots it may be briefly blinded by the Sun, but only for a matter of seconds; the modest inferno at the front of that curving, elongated serpent dramatically increases in brightness and size until it's all anybody looking up can see. The atmosphere does its best to capture the conflagration, and just thirty-five miles above the Earth's surface, the projectile does begin to crack and shatter. But it's already much too late. It's too structurally sound, big, and fast to be stopped. For a few heartbeats, as the asteroid is ever-so-slightly slowed down and fragmented in the skies above Seattle, an enormous amount of energy erupts outward in every direction. And just three seconds later, at precisely 8:23 a.m., somewhere close to the corner of Pine Street and Eighth Avenue, a ruptured 460-foot asteroid that punched through Earth's atmosphere at 40,000 miles per hour comes to a complete stop.

The explosion at the heart of the impact site defies description. Our planet barely notices it has been hit; to that ancient mass of rock, metal, and time, it feels like a pebble being shot into a vast and unperturbed beach. But the tiny humans at ground zero see their lives end in a luminous and nightmarish instant; what once was solid is now vapor. The unfathomable momentum of the asteroid and the kinetic energy of the impact—equivalent to several million tons of TNT exploding all at once—has turned crystalline rock, concrete, and tarmac into a fluid, carving out a temporary, amorphous hole more than 4,000 feet wide and 1,500 feet deep in the blink of an eye.[1]

There isn't a giant fireball; this asteroid was too puny, and hit Earth too slowly, to vaporize enough matter to turn everything aflame. But its violent hello creates quakes so intense that the entire city vibrates. If you were standing just far enough away to avoid being swallowed up by the impact's ephemeral maw, it only takes three or four seconds for you to be hit by a compressed wall of air moving up to 9,000 miles per hour. If your eardrums manage to survive the primeval roar, you may see, just for a moment, a litany of destruction: trees lining the streets are uprooted and flung into the air as if made of paper; cars are wrapped around lampposts like warm candle wax; windows turn into a horizontal maelstrom of glass as skyscrapers and office blocks get knocked back and comprehensively demolished.

If, by some miracle, anyone on the edge of the transforming crater manages to survive the airburst, they will then look up to see a tsunami of rocks, roads, and buildings obfuscating that once-blue, beautiful sky. Just fourteen seconds after impact, a monstrous swarm of asteroidal and terrestrial debris, composed of fragments ranging in size from boulders to entire city blocks, overwhelms and consumes these people so quickly that, mercifully, their nerve endings do not get a chance to register pain.

Anyone a few miles from ground zero may survive. But from the time you started reading this paragraph to the moment you finish this sentence, the city's heart would have been carved out; the briefly liquid-like earth has now settled, leaving a lasting crater a mile across and 1,100 feet deep in its place. The air blast would have taken out thousands of buildings, and the falling curtain of ejected debris would have flattened or eviscerated much of the surrounding metropolitan area. Forget, for a moment, the chaos, the fires, the peripheral and unpredictable damage such an impact would cause. The rendezvous between that asteroid and the city that, by chance and chance alone, was unfortunate enough to get in its way, immediately ended the lives of a million people.

To the rest of the world, it would be like an entire city had been deleted. Phone calls would suddenly be severed. Text conversations would abruptly stop. Most communications out of Seattle would mysteriously drop off. But it wouldn't take long for everyone to realize the horror that had unfolded.

By 2046, thousands upon thousands of asteroids just like this one—similarly sized rocks flitting about in Earth's neighborhood, those capable of devastating a city or ruining an entire country—have been spied by humanity's multitude of astronomical eyes over the past half-century. Our cosmic cartographers plotted out their orbital journeys around the Sun, and none of them stood a chance at hitting Earth for at least a hundred years into the future. But not this one. This one, through horrific serendipity, was destined to destroy Seattle the moment it was freed from its asteroidal parents 10 million years prior. Like many other comparatively small asteroids, it had not been found by astronomers. The next-generation observatories that could find almost all these meteoritic missiles were deployed a little too late. And even if astronomers had spotted this asteroid years before impact, they would have been helpless to do anything about it. Earth had no effective way to defend itself from this cosmic catastrophe. We had been gambling with the universe for too long, and on February 19, we were dealt a bankrupting hand. It was inevitable: during this lifetime, or during one of the many lifetimes to come, hundreds of thousands of people—perhaps millions, potentially tens or even hundreds of millions—were always going to die. And so, on that sunny Monday morning, they did.

This hasn't happened, of course. But it could. If we wait long enough, an asteroid like our Seattle killer will hit Earth. That's not a possibility, but an absolute certainty. If we're extremely fortunate, it could land in the ocean, or in the middle of an expansive and uninhabited desert. But if it strikes close to a major metropolitan city, or smack bang in the middle of it, it will prove to be the single worst disaster in human history.

Some astronomers have given asteroids with dimensions close to 460 feet a catchy and appropriately ominous moniker: city killers. The unnerving truth of the matter is that an asteroid half that size could handily destroy a city. Double that size, and an entire US state, or a small nation, could be wrecked. Triple it, and most countries wouldn't survive. Forget the planet killers, asteroids at least two-thirds of a mile long: thanks to their mammoth dimensions, astronomers have found almost all—if not quite every single one—of these civilization enders. Their orbits don't take them perilously close to Earth for at least a century, if not far longer. But scientists know that there are tens of thousands of city killers not especially far from all 7 billion of us—and, at the time of writing, most of them are yet to be found.

The good news is that the day-to-day odds of any of us experiencing a horrendous high five between Earth and one of these unwelcome nomads is extremely low. Over a lifetime, those odds remain low enough that you should concern yourself far more with everyday risks, from getting pancaked by a speeding car you failed to see while crossing the road to getting struck by a particularly mean-spirited bolt of lightning. But the fact remains that, one day, a city killer will strike our planet—and the odds that you will live to see it happen are higher than anyone would prefer. All of us are potentially at risk, as are any of our descendants. Long ago, in the unimaginably distant past, the baroque gravitational dance of would-be worlds began a chain of events that will result in a death-dealing asteroid impact at some point in the future. The question, then, is this: How lucky do you feel?

But unlike any other disaster of natural origin, this one has an exceedingly welcome caveat: we can prevent it from happening. Normally, damages and death can merely be mitigated. Science and technology, with the right people in power, can provide forms of armor to those at risk, or can apply gauze and bandages to those wounded by geologic or meteorologic violence. But nobody can actually stop an earthquake. Nobody can halt a volcano from erupting. Nobody can prevent a hurricane from taking shape and annihilating coastlines. It is scientifically possible, however, to make sure an asteroid does not strike Earth. Within the next few decades, if the right decisions are made and those in the know get the chance to prove their worth, something that once belonged stubbornly in the realm of the imagination can become a perfectly ordinary part of reality. In such a future, most, if not all, of the city killer asteroids in humanity's galactic backwater will have been found—and any determined to be on a collision course with Earth will be forever prevented from making that meeting. An entire type of disaster will be taken off the board, wiped from our ever-growing list of existential dreads. We will have made the planet more habitable by subverting a tragedy before it ever gets the chance to transpire. Those millions fated to perish, to be mourned and grieved by their loved ones, will get to live, unaware that the unthinking universe had its macabre plans spoiled by human ingenuity, charity, and proactivity.

We are, all of us, living through this very inflection point in human history. After decades of persistence, an international team of maverick scientists and engineers are hoping to make that science fiction–sounding ideal a fact. They have long dreamed that Earth would one day have an effective planetary defense system—and the realization of that dream all hinges on the dramatic events of a single day, as a spacecraft 7 million miles from home heads toward its grave.

It's September 26, 2022, in Laurel, Maryland, a short drive north of Washington, DC. I'm inside a huge room on the campus of the Johns Hopkins University Applied Physics Laboratory, watching an asteroid loom ever larger on a giant screen. I'm joined by hundreds of other people—astronomers, astrophysicists, young and old, all deathly quiet. The cacophony that filled the entire campus seconds earlier has given way to fading echoes, almost silence. Every now and then, the stilted air is pierced with an expletive, or a sharp intake of breath.

The screens are displaying close to real-time images of an asteroid and its moonlet, another asteroid captured an age ago by its larger sister's gravitational well. We are seeing through a robot's eye: the single optical camera on a winged spacecraft heading straight for that moonlet. What was a collection of gray pixels mere minutes ago is now a world of its own, a boulder-adorned creature of dangerous dimensions. An asteroid this size could, should it visit Earth, flatten any megalopolis. If left alone, such an asteroid will fulfill this astronomic prophecy. And so, for these scientists and engineers, what needs to be done before that day arrives is painfully obvious: pilot an uncrewed spacecraft toward that asteroid and crash straight into it, hitting it with enough oomph to change its orbital path around the Sun.

And so here we are. On this September evening, the apotheosis of decades of work by countless astronomers rests in the hands of just a few dozen key individuals. A spacecraft the size of a car is speeding through the vacuum of space in the hopes of hitting a bullseye moving tens of thousands of miles per hour. It feels like an impossible shot. Will it make its mark? If they pull this off, it will mean that humanity has the capability to defend itself, to deflect a future asteroid that might be, unlike this target, genuinely coming to kill us.

As the asteroid blooms on the theatrical screen before us, I can tell that the excitement, terror, and abject absurdity of this deep space, smashy crashy experiment is inundating the spacecraft's human creators. Its scientists are holding their breath. Its engineers, eyes unblinking, are transfixed. Across the world, and elsewhere in space, dozens of telescopes are trained on a tiny speck in the void. And inside the mission control room, the spacecraft's commander is clasping her hands together as the asteroid becomes all the lonely ship can see. This is it: a chance to alter the course of not just an asteroid, but human history itself.

This book is not just a story about science. It's about a small group of scientists and engineers doing what they can to save millions of lives from a future catastrophe.

And this is how that story begins.

HOW TO KILL AN ASTEROID

1

To Russia with Love

Back in June 1994, Major Lindley Johnson had a thought: Wouldn't it be nice to try to save the world? It sounded perfectly reasonable to him. But to everyone else it seemed deeply silly.

A year earlier, the US Air Force's chief of staff had asked Alabama's Air University—a military college for those keen on defying the conventions of gravity—to peer into a crystal ball and take notes. This commissioned report asked various experts to ruminate on the challenges American warfighters would face in the twenty-first century, and then offer technological ways to meet those challenges. Known as SPACECAST 2020, the document was a smorgasbord of post–Cold War wishful thinking—how to achieve supremacy in space, ways to use the flow of information to project soft power, and so on—and featured ponderings by military officials, members of the intelligence community, scientists, engineers, Arthur C. Clarke, Carl Sagan, and the screenwriter for *Terminator 2*.

The most important contributor to this report (at least for our tale of world-saving aspirations) was Air University's Lindley Johnson. Having graduated from the University of Kansas with a bachelor's in astronomy, he was already an atypical member of the Air Force—and his white paper for SPACECAST 2020 did nothing to dissuade his colleagues of that. The title? "Preparing for Planetary Defense."[1]

Earth, Johnson boldly noted, was in the firing line. Plenty of asteroids are out there, and plenty cross Earth's orbit around the Sun. Earth has been hit many times in the past, sometimes cataclysmically. It'll be hit again, and our atmospheric shield will only stop the most Lilliputian lithic missiles. A large asteroid could kill billions and destabilize or topple civilization as we know it; a small asteroid could maim millions. Both events will happen someday, no matter how much we hope it won't. "The probability is finite. Indeed, one day it will be exactly equal to one," he wrote. "Currently, astronomers have no idea when that day will come."

When we first spoke, Johnson bluntly outlined why America should care, even if America isn't the target of the next cosmic bombardment: "National security is all about preventing and avoiding situations that could destabilize governments; not just our governments, but [those] of our allies," he explained—meaning that asteroid impacts are a national security concern. His suggestion for addressing this concern? Work out how to turn missile delivery systems and nuclear warheads into tools of planetwide salvation, while watching the skies closely for any natural objects that may pose a threat. He also suggested setting up an Air Force office to run the show.

Much like his conversational style, Johnson's written prose is concise and considered. Despite the scale of the looming global problem, his report managed to condense the issue down to a few pages, making the key point—we should really do something about this *right now*—abundantly clear to the powers that be. What did the Air Force think of it? He chuckled to himself over our Zoom call. "You know what they thought of it," he said with a ceiling-ward look. "I got laughed at."

The idea of "planetary defense" was hardly original. The problem, perhaps, was that it was not perceived to be a grounded, militaristic matter, but a very different sort of enterprise. "It's kind of been a mainstay of science fiction for quite a while, both books and movies," he said. "In fact, there was a *Star Trek* episode that was very much about planetary defense: 'The Paradise Syndrome' episode."

I jumped onto Google. This episode—the third of season three of the original *Star Trek* television series, first broadcast on October 4, 1968—features a doomed love story and a bout of amnesia for Captain Kirk. It also, more pertinently, involves the crew of the USS *Enterprise* trying to stop an asteroid slamming into an alien planet and killing everyone. They try using the ship's deflectors at first, but it's not enough, at which point they attempt in vain to cleave the asteroid into bits with their phasers. Ultimately, Kirk, Spock, and McCoy team up on the planet's surface to activate an alien world's ancient asteroid deflector, which successfully flicks the projectile back into the depths of space.[2]

Specific plot points aside, lofty notions of saving millions of people or entire worlds fall predominantly within the purview of superheroes and space-based swashbucklers, not members of the Air Force. This perhaps excludes the movie *Independence Day*, but I'd wager that Johnson never planned on killing Earthbound asteroids by gleefully declaring "Hello boys, I'm back!" as he crashed his warplane into the rocky beast's underbelly. Either way, it's not difficult to see why the upper echelons of the Air Force of the mid 1990s took one look at Johnson's proposal and raised their collective eyebrow in bemusement. Why should they have to worry about anything crashing into Earth when they had hostile nations and state actors to worry about? At the same time, the sci-fi trappings of a term like planetary defense can provoke the wrong impression. "I know for a lot of them, aliens leap to mind," he said.

Then, in July 1994, the universe gifted Johnson with one of the most simultaneously gratifying and unnerving "I told you so" moments of all time. That year, a comet big enough to kill not millions, but *billions* of people, crashed into Jupiter. "And suddenly, cosmic impacts became a reality," he told me. "That kind of saved my thesis topic, because all of a sudden they said: Ah, well, maybe we *do* want to hear about this."

Any American astronomer worth their salt has, even in passing, heard of the late, great Shoemakers.

Gene, the husband, was a gifted geologist. While working for the US Geological Survey in the early days of the Cold War, he began to wonder if those great big holes on the lunar surface were caused by volcanic explosions—the prevailing thought at the time—or by asteroids and comets smashing into the Moon. While mapping nuclear bomb craters, he noted how similar they looked to not just the circular crevasses on the Moon, but those on Earth, too. Eventually taking on the mantle of "astrogeologist," he became convinced, through the extensive examination of craters familiar and alien, that most of these pits were caused by cosmic collisions and not magma-filled mountains. That made him realize that Earth gets hit by these extraterrestrial interlopers more commonly than most people would prefer. Medical troubles prevented him from becoming an Apollo astronaut, but he spent the rest of his days forensically examining Earth's battle scars while kickstarting programs to search the skies for near-Earth asteroids and comets.[3]

Carolyn, the wife, obtained a master's degree in history and political science in her youth—but her marriage to Gene, and his own curious profession, convinced her to switch academic alliances. After the couple raised their three children, she decided at the age of fifty-one to do something completely different: study the stars and the rocky and icy ships that sail between them. In 1982, Gene and Carolyn founded the Palomar Asteroid and Comet Survey, a California mountain-based telescopic effort to catch these primordial remnants dancing about. Today, scores of asteroids and dozens of comets bear the surname of their eponymous discoverers, along with those of some coconspiring astronomers that joined their search party.[4]

By 1993, Gene and Carolyn were already legends in the astronomy community. But what they found shimmering in space that March indelibly inked their names in the history books. Along with astronomer David Levy, the Shoemakers had spent several days that month atop Palomar Mountain, grumbling about the crappy weather and subpar photographic film. During the night of March 18, for just a moment, the clouds parted, like the foam at the tip of a tide being washed away, and they photographed that patch of inky-blue night on their imperfect film. Later, during another overcast and rainy night, Carolyn was perusing their clear-sky images. Nothing, nothing, nothing . . . wait. A smeared, fuzzy entity could be seen not far from Jupiter.[5] Gene popped over and had a peek. So did David. They went to wake Philippe Bendjoya—a visiting French astronomer who was taking a nap outside in a car—who also had a good, hard look. And they all collectively agreed that they didn't quite know what they were looking at.[6]

The smudge resembled a comet, but not *quite*. It almost looked like lots of small comets traveling together, like a herd of pack animals or a flock of migrating birds. Gene suggested it might be a "broken comet"[7]—in other words, a recently shattered orb of ancient ice in its death throes, now appearing as a string of gelid pearls. And he was right: by March 23, other astronomers had used different telescopes to look themselves, and they had indeed discovered a comet that had fallen apart. It was christened Shoemaker-Levy 9, being the Shoemakers' and Levy's ninth co-discovery.[8]

That the comet, once roughly a mile wide,[9] was found not long after it had splintered—perhaps sometime in early July 1992, when it got a little too close to Jupiter and was torn apart by the tides of the gas giant's intense gravity—was certainly interesting. But what made everyone's hair stand on end was the fact that these twenty-one fragments were orbiting Jupiter, not the Sun—and that, in the summer of 1994, this corpse of a comet was going to repeatedly plunge into Jovian skies.

Almost immediately, the astronomical community rallied. Every available Earth-based telescope, the Hubble Space Telescope, and even a couple of uncrewed spacecraft, such as the Jupiter-bound Galileo mission, turned to face Shoemaker-Levy 9, watching eagerly as the frigid shards, each with their individual tails, tumbled gracefully toward the solar system's heftiest planet. By May 1994, it had been calculated that the fragments would individually shoot into Jupiter's stormy carapace between July 16 and 22.

Media attention ramped up as the infernal week approached. During a lengthy PBS special,[10] Carolyn Shoemaker was asked how she would feel if a comet with her name attached to it were heading for Earth. "I think I might want it named after someone else," she replied, before adding: "But then it wouldn't matter too much." Fair enough: if this comet had instead hit Earth, it would have killed billions of people, and the plume of dust created from the impact would have stood a good chance of flooding into the higher layers of the atmosphere, blocking out sunlight for years or decades. That would have prevented photosynthesis, decimated the world's plant population, and caused food chains to collapse. Life would no doubt have found a way to survive, but Earth and its ecosystems would be irreversibly transformed.

That Shoemaker-Levy 9 was going to instead impact Jupiter, a lifeless world so gargantuan (1,321 Earths could fit inside it[11]) that the collisions would be as inconsequential as flies hitting a car's windshield, delighted everyone. Well, almost everyone. Carolyn was used to discovering comets that would safely orbit the Sun for eons into the future. This comet was the first she had found that had an imminent date with death. "We are sorry to see it die," she said during that PBS special. "Well, it's going to make a big bang. We'll be happy with that."[12]

On July 16, the meteor shower to end all meteor showers began. The first fragment, moving at 133,200 miles per hour—fast enough to go from New York City to Los Angeles in just a single minute—hit Jupiter, piercing through the tempest like a bullet through butter. Instantly, the normally chilly skies burned at 53,000 degrees Fahrenheit as high-altitude clouds rushed and cascaded across the planet in every direction. Moments later, a tremendous pillar of debris shot nearly 2,000 miles above the Van Gogh–like cloud tops, concealing a blackened, bubbling scar below.[13]

And this happened again. And again. And again. Fragment G, the largest of the twenty-one pieces, struck on July 18. It produced a flash so luminous that telescopes were overloaded and unable to measure its empyreal brightness; its impact zone was roughly equivalent to the size of Earth. The entire spectacle, which made the front page of the *New York Times*,[14] reportedly unleashed the energy of 300 million nuclear bombs.[15] That's almost impossible for anyone to visualize in any meaningful or accurate way.

But Kelly Fast didn't have to try. She was there.

"I was a young astronomer at the time," she told me. Back in the mid-1990s, she was working as a contractor at NASA's Goddard Space Flight Center in Greenbelt, Maryland. She was studying the elements that made up the atmospheric soup of various planets—and it turned out that one such world, and its atmosphere, was about to get plenty agitated. As the dazzling images of the Shoemaker-Levy 9 impacts flooded in, she found herself lost for words. When I spoke to her, she had had almost three decades to ruminate on what she had seen, and upon asking what the experience had been like, I expected her recollection of the explosive events to be a stream of superlatives. Instead, she took a deep breath, grinned, and said, simply: "It was wonderful."

I had arrived in Washington, DC, in the fading summer months of 2022, and as part of a months-long journalistic mission, I had stopped by NASA's headquarters: a relatively unassuming collection of boffin offices adorned with the requisite display of American flags. Just outside the entrance, a sticker adhered mischievously to a column protested: Pluto Is STILL a Planet! One of my objectives was to finally—after chatting on plenty of pandemic-era Zoom calls—meet Fast, a perpetually enthused, ebullient personality and a key player in the planetary defense community who had been inspired by fantastical and futuristic fables.

"I'm from Hollywood," she told me. "I was such a geeky kid." From her childhood home, she could see both the Hollywood sign and the Griffith Observatory—setting the stage for an alchemic mixture of science fact and fiction. "We had family friends who were actors." Was she ever tempted by the limelight herself? "I was an awkward geeky kid, off in my own world. I loved *Star Trek*. I saw Walter Koenig in the post office once and I was too embarrassed to go up and say: 'I loved you when I was a kid!'" Her pop culture credentials, already extremely strong, broke through a critical ceiling when she said, rather casually: "I saw *Star Wars* when it opened at the Chinese Theatre."

The adventure of space handily triumphed over Fast's other temporary childhood career dreams. Although allured by the academic promises of distant galaxies, like many astronomers, the study of our own solar system ultimately won out for her. "I just think we can relate to it more. It's something tangible," she said. The frankly irritating expansiveness of the universe means that nobody will ever live to reach another galaxy. But Mars? Jupiter? Even (dwarf) planet Pluto, way out in the incomprehensible darkness? "You can send robotic spacecraft to look at them."

That notion of tangibility, and one of impermanence, hit home in the summer of 1994. Those nearly two-dozen thunderous blows bruised, but did not permanently scar, Jupiter. But replace that vaporous titan with Earth, and you've got yourself a planet killer. Scientists knew that impacts have happened, and will always happen, in the solar system. Lindley Johnson, writing his planetary defense thesis for SPACECAST 2020, made it clear that in another timeline, that comet, or an asteroid, could have hit Earth that July. But nobody took him seriously. "And then Shoemaker-Levy came along and put a bunch of exclamation marks on it," said Fast. Until this point, the violent reality of asteroid and comet impacts only existed in the collective imagination of astronomers, a mental jigsaw made of pieces extracted from geologic wreckage and ruins. Finally, they had front row seats to the fireworks.

Asteroids and comets are 4.6-billion-year-old balls of rock, metal, ice, and dust from the beginning of the solar system—they are like corporeal memories, dating back to and sometimes predating the birth of the Sun. They tell the story of the stars and worlds around us, including our own. And that's fantastic. But this is a book about defending Earth from the comets and asteroids that have decided it's time for an apocalyptic meet and greet. That means three things matter to us: what asteroids and comets are made from, where they are made, and how they find their way to us.

Observed by humans since time immemorial, comets have been varyingly considered all sorts of things, from wandering planets to harbingers of doom to strange deific hairy orbs; their name stems from the Greek *kometes*, which means "hairy one."[16] And although the scientific definition of what makes a comet is becoming rather fluid, its general anatomy is as follows.[17] Its core, consisting of various ices—not just water-ice, but frozen carbon dioxide, carbon monoxide, methane, and/or ammonia—and a bunch of other molecules, is named the nucleus. When, during its circumnavigation of the Sun, a comet begins to make its closest approach to that stellar furnace, its most volatile ices begin to vaporize into a gas, producing a hazy, reflective, ephemeral atmosphere called a coma. This aggressive release of gas jettisons dust into space, whereupon it is hit by photons—the particles of sunlight—and pushed out into a curved tail. The solar wind, an electromagnetic soup continually belched out by the Sun, also manages to excite various particles flitting off the active comet, dragging them out into space and forming a second tail.[18]

Comets have two cradles.[19] The first, the Kuiper belt—named after astronomer Gerard Kuiper—can be found just beyond the distant orbital kingdom of Neptune. It's a donut-shaped torus of primordial icy chunks that failed to coalesce into bigger things, like proper planets, but remained sufficiently distant from the Sun to stay frozen for billions of years. Pluto is one such Kuiper belt object, but there are plenty of others, such as Arrokoth,[20] which looks like an incomplete snowman put together by an inebriated god. All these objects—seen mostly through telescopes and space observatories, but infrequently gawked at by passing spacecraft[21]—orbit the Sun. Those that get close enough to the Sun to develop a coma become short-period comets, the sort that take less than 200 years to go once around the solar system velodrome.

The alternative and likely far more populous comet factory is the Oort cloud,[22] named after astronomer Jan Oort. Hypothetically, this is a stupendous sphere of icy objects, all unfathomably far from one another, ranging from the size of cats and cars to mountains and misshapen moons. Nobody has ever seen it, though, because the inner edge of the cloud is thought to be as much as 5,000 times the distance between Earth and the Sun. Every now and then, a long-form comet, which takes anywhere between 200 years and several tens of millions of years, perhaps more, to orbit our local star, pops out of this spooky, Lovecraftian realm.

That's pretty much all there is to comets. When they burst to life as they approach the Sun, they are unquestionably beautiful. But that beauty turns bestial when you imagine one of them hitting Earth. Why, you may wonder, would they do something to ruin such a stellar reputation?

You would do just as well to ask the same question of asteroids. The history of these "starlike" objects, per their etymological origins, is fascinating,[23] but can be summed up thusly: in the early nineteenth century, astronomers found some truly large and fast-moving rocks in the night sky. They found only really big ones—some the size of small planets—until they got better telescopes. Then they started finding smaller ones. At the time of writing, the known asteroid count is upward of 1.3 million.[24] Most are made of rock, some are filled with metal, and they sometimes have ice on them. And almost all of them—but, crucially, not all—orbit the Sun in the gap between Mars and Jupiter in the aptly named asteroid belt.[25]

What's the deal here? Why, among these neatly ordered planets, is there this huge dumping ground of rocky litter? Why aren't they all hanging out with the cool kids in the Oort cloud? And what's this got to do with saving the world? Here's Gretchen Benedix, an asteroid expert at Curtin University in Perth, Australia, to explain more:

"I call myself an astrogeologist," she told me. "I study rocks from space." Regular rocks from Earth? Shrug-worthy. Ones that zip about at tens of thousands of miles per hour? Yes, please.

So where did those 1.3 million asteroids come from? "That's a really, really good question," she said. (That's what scientists say when nobody is entirely sure.) "Maybe the asteroid belt represented a planet that blew up." Okay, very cool, very cool. A world between Mars and Jupiter got blown up by the astronomical equivalent of the Death Star, which, in this case, was another planet-size object, or a few of them, crashing into a still-forming embryonic planet, shattering it into more than a million pieces.

But the real villain here may have been Jupiter. Back in the fire and brimstone era of the solar system, when the planets were still self-assembling, the emergence of this swaggering gas giant threw a massive gravitational wrench into the works. Venus and Earth are distant from Jupiter, and are pretty much the same size. But Mars is about one-sixth of that volume, and closer to Jupiter. Scientific simulations suggest that Jupiter may have eaten up a decent chunk of its building materials—and when weighty Saturn, Uranus, and Neptune entered the fray, the subsequent tug of war ensured that any remaining planetary concrete was fixed in place by their gravitational tendrils.

Asteroids—between Mars and Jupiter. Comets—behind Neptune, mostly. Why do some of them end up striking planets, including our own? Seems rude. That is also down to gravity's eccentric embrace. The solar system may look ordered on a diagram. And for the most part, it is. The planets stick to their slightly elliptical orbits. But their gravitational influence—particularly that of Jupiter, whose prodigious mass creates a truly ginormous gravitational well—gradually tinkers with the voyage of any comet or asteroid that passes a little too close. Add up those miniscule pulls over billions of years, and eventually some of those comets will arc toward worlds, while asteroids get pulled into the solar system's inner sanctum, far from the belt, where they can potentially cross Earth's own highway.

At this point, it's starting to sound like Jupiter deserved to have Shoemaker-Levy 9 crash into it. "Mm. It is a bit of a bully," conceded Benedix.

Comets, but particularly asteroids, can also fashion their own rogue shards.[26] First, though, I must deliver a crushing disappointment. As a chuckling Benedix noted, the asteroid belt is not like the one in *The Empire Strikes Back*. Asteroids are almost never next to one another. They are hundreds, thousands, sometimes hundreds of thousands of miles apart. (In our solar system, Han Solo, trying to escape those pursuing TIE fighters, would have had no such rambunctious melee of space rocks to weave through.) Nevertheless, Benedix pointed out, "Even though there is a lot of space out there, they will occasionally bump into each other." Boom. Bits go everywhere. Some of those bits hit planets, and if the impacts are big enough, some of that planetary scar tissue makes it back into space, magically transforming into asteroids themselves.

Over the last 4.6 billion years, Earth has been hit by a fair number of comets, which tend to be problematically huge. But all those asteroids lingering closer to home, so easily led astray by collisions and gravity's whims, means that far more of those rocky rebels have pierced through our planet's thin atmospheric bubble. So, despite the fragments of Shoemaker-Levy 9 demonstrating their horrifyingly destructive potential, comets are the deuteragonists in our world-saving tale. To you and me, asteroids are our chief antagonists.

One last thing: a note on nomenclature. "Rocks in space: asteroids," said Benedix. "Rocks on ground: meteorites." Asteroids entering Earth's skies? Meteors. And if those meteors are big enough? Death.

Lindley Johnson was not the first astronomer to think about planetary defense. "I was kind of there close to the beginning," explained Clark Chapman, a planetary scientist at the Southwest Research Institute in Boulder, Colorado, and one of the field's first proponents. "Other than in science fiction, the real story began in the 1980s, here in Colorado." In 1981, NASA asked Gene Shoemaker to herd a group of scientists and rocket engineers into the town of Snowmass to discuss a cheery topic: How often does Earth get hit by asteroids, and what, if anything, can be done to stop a future face-off?

The group took the subject matter deadly seriously. But, as Johnson would experience a decade later, nobody else did. " 'Rocks falling from the sky? What the heck was all that about?' " Chapman said, echoing the wider community's incredulity at the time. Fortunately, the Snowmass group's eschatological musings gained much-needed credibility when some familial sleuths claimed to have solved an epic whodunnit: Who, or what, killed the dinosaurs?

You're reading a book about killer asteroids, so I'm going to safely assume a spoiler warning isn't needed here. Most scientists agree that 66 million years ago, a six-mile-wide asteroid slammed into what is now Mexico's Yucatán Peninsula, and Earth had a goddamn awful day.[27] Even before the asteroid made contact with the ground, the compressed air in front of it would have sizzled at temperatures greater than that of the Sun's surface.[28] Upon impact, a hellish fireball would have incinerated anything on the same side of the horizon as the impact. Like firing a cannonball at a pane of glass, unimaginable earthquakes would have effortlessly sundered parts of the gigantic continent. A tsunami over a mile high would have washed over[29] much of the world. Most devastatingly, the sheer volume of debris and dust propelled into the upper atmosphere stayed up there for years, blotting out the Sun and ushering an age of climatic, environmental, and ecological devastation that murdered three of every four species.[30]

Despite being 110 miles wide, the planet killer's impact crater became largely submerged beneath the Gulf of Mexico as the tectonic plates continued their inexorable march. Paleontologists, keen to know what killed those often-giant reptilian beasts, found revelations in a scientific paper[31] published just one year prior to the Snowmass gathering. Physicist Luis Walter Alvarez and his geoscientist son, Walter, along with some of their colleagues, had been digging into a 66-million-year-old rock layer that could be found all over the world. No matter the locale, this layer contained plenty of iridium, an element found infrequently on Earth but in abundance in asteroids. The Alvarez team realized they had struck scientific gold. If an asteroid impact did in fact scatter iridium across the globe, then such a collision would be the prime suspect in the case of the slaughtered dinosaurs.

Unbeknownst to them, the smoking gun—the crater itself—had been found two years earlier by Glen Penfield, a geophysicist working for Pemex, a Mexican petroleum company, while scanning the seafloor near the Yucatán Peninsula. But Penfield's discovery was deemed unimportant by his superiors and went unnoticed by the wider scientific community until 1990, when a reporter directed planetary scientist Alan Hildebrand to look at Penfield's seafloor charts. In 1991, Hildebrand, Penfield, and other colleagues published a paper[32] describing the 66-million-year-old crater beneath the sea. The scar, dubbed Chicxulub, was ground zero for the Alvarezes' dinosaur-killing mass extinction event.

"The reality of asteroids being a danger and having been a danger to the Earth in the past—you could attribute that to the Alvarez paper," said Chapman. Shoemaker-Levy 9's collision with Jupiter provided a contemporaneous spectacle reinforcing the notion that Earth is wholly unprotected from this threat.

The Shoemaker-Levy 9 impact is thought to have partly inspired Hollywood to create the 1998 planet-killer double bill of *Armageddon* and *Deep Impact*. But it provoked action in the real world, too: NASA began to track all newly identified asteroids and comets, just in case one looked like it was about to ambush us. And by the mid-1990s, for planetary defense proponents, "the giggle factor had largely gone away," said Chapman.

And yet, it remained a largely unfunded and unfocused effort. Space agencies did not make it anything close to a priority. Even by the early 2000s, Chapman said, "NASA just wasn't interested. Nor was the European Space Agency, or the Russians." Lindley Johnson remained zealous about his thesis, and his dreams of setting up an American agency that aimed to save the planet. But his voice was drowned out.

Until, that is, an asteroid crashed into Russia.

In one very specific way, an astronaut in a spacecraft is not too different from an asteroid: both aspire to make it to the ground in one piece. In February 2013, the ways in which both things try to survive a dive into Earth's atmosphere were the topic of a scientific conference in California—and Peter Jenniskens and Olga Popova had Russia on their minds. It's a huge landmass. Asteroids, statistically, should impact it more than other countries. And in their lifetimes they hoped to see one smash into the former Soviet Union.

Not in a malicious manner, mind you. Jenniskens, of the California-based SETI Institute (which, among its other goals, hopes to find a transmission in space sent by alien life), is known to the community as an expert in meteorites and finding them wherever they may fall. Popova, of the Moscow branch of the Russian Academy of Sciences, is an expert in why things from space smash into other things. Together, they thought it wouldn't be the worst thing if a tiny asteroid burned up in the skies above Siberia and parts of it landed somewhere where they could be retrieved. And on the sidelines of this conference, Jenniskens pondered how such an expedition would operate in the notoriously authoritarian state.

"And then one week later, there was news of a 500-kiloton explosion from an asteroid impact over Chelyabinsk," said Jenniskens. "What are the odds?"

On February 15, at 9:20 a.m. local time, the 1.2 million people living in the Russian city of Chelyabinsk went about their Friday, perfectly unaware that the firmament was about to rudely interrupt them. About sixty miles above the snow-dusted earth, something entered Earth's atmosphere.[33] As seen in footage from dozens of dashboard camcorders and a smattering of CCTV cameras,[34] many in the city were initially stunned to see the ground in front of them light up, as if illuminated by a spotlight with an out-of-control dimmer switch. They looked up, and a streak of light, trailed by dust and flame, was arcing through the sky at sixty times the speed of sound, heading directly toward them. Just fifteen seconds after it fell into Russian skies, its embers erupted. A flash. An explosion. The white light on the ground flipped to a blood-red hue. A punch to the head and chest. A deep, guttural banshee howl. Within microseconds, windows exploded inward, creating shotgun blasts of glass. Those outside and close to the blast were knocked flat on their back or swept off their feet. Roofs collapsed as walls cracked, but just about held their ground. The incandescent light, quickly snuffed out, gave way to several glowing fragments that fell across the region, landing at sites unknown—the split, cremated corpse of an asteroid that failed to land, leaving an acrid odor in its wake.

"I had a phone call from my colleagues, who said there is some... event in Chelyabinsk," Popova recalled. As reports of an explosion over Russia began to circulate in the media, she arrived at her office in Moscow and saw the dashcam footage. "Then, I understood." Her wish had been granted—but this asteroid was far from pint-size. It looked like people had been hurt, perhaps even killed.

"Very quickly, we had a team assembled," said Jenniskens. He went to a Russian embassy and nabbed a visa. "Three weeks from the event, I found myself in an aircraft flying from Moscow to Chelyabinsk." When he landed, the cold chewed at him through his clothes—and before he had time to adjust to the cold climate, he and Popova were sat down in front of the region's governor. "At the end of the table was a battery of television cameras," he said. Nobody really knew what to say—after all, it's not every day your city gets broken by an asteroid's self-destruction—so Popova explained what they were there to do, and the governor nodded enthusiastically, before offering them a car and a driver.

First, they needed to evade the press. "Many TV companies wanted to follow us on the expedition. It was too much for us," said Popova. "We even escaped from one TV company. We came from one door and left through another." After eluding the meteor-mad media, the pair were off on an expedition like none other, with two objectives: reconstruct the manner in which the asteroid perished and find any surviving fragments. The latter was accomplished quite handily: pieces were found early on, often by residents of the city and the wider region; not long afterward, aided by the recorded footage of the meteor, a hefty chunk weighing two-thirds of a ton was found at the bottom of a shallow, iced-over lake with a brand-new hole in it.[35]

Reverse engineering the events of the asteroid's death was a singularly strange experience. Reconstructing the explosion meant determining the blast radius of infrastructural damage and human injuries. And Jenniskens and Popova didn't just venture around Chelyabinsk; they ended up leaving the city entirely, driving between rows of frost-flecked fields and diamond-dusted trees to reach villages claiming to have eyewitnesses to the meteor.

Coincidentally, the area was familiar with unfortunate explosions. At one point, Jenniskens and Popova were quickly escorted around the periphery of a radioactive swamp, one haunted by thousands of ghosts. Ozersk, a locale code-named City 40 by the Soviets, is just a few miles northwest of Chelyabinsk. There, in the 1940s, a nuclear reactor was built to produce enough plutonium for the country's first atomic bombs. They had succeeded by 1949, but according to a report in the magazine *Nature*,[36] the well-worn apparatus had begun to crack open, leading authorities to ask workers to shift tens of thousands of radioactive tubes by hand. Mass radiation poisoning, and plenty of death, ensued. The nearby river, whose banks hosted several towns, became irradiated. And in 1957, a subterranean nuclear waste container exploded.

Like all asteroids, 2013's visitor delivered no radioactive matter to the surface. But it did unleash an otherworldly fury. In total, Jenniskens and Popova visited fifty different villages, some as far as ninety miles from Chelyabinsk. And at almost every destination, they found lesions left by the interplanetary invader.

"People were really impressed by the event. They remember many details," Popova recalled. Immediately after the explosion, some initially assumed they were under attack, and that a missile was to blame. After quickly establishing that that wasn't the case, everyone reacted to the blast differently—and that mix of emotions was still on full display by the time Popova arrived on the scene. "Some people said, 'Ooh, it's wonderful.' Others said. . . ." She dramatically gasped and held her hands up, laughing a little. They were afraid. And those injured? Naturally, "they would prefer it not to happen again."

At a distance, the explosion sounded "not like a sharp shock, but more of an extended drum roll," said Jenniskens. Although glass panes had been broken in villages far from the city, Chelyabinsk itself was the most severely afflicted. A massive bang coincided with a shockwave thundering into the metropolis; in 3,613 apartment buildings, "windows turned into a spray of glass" while doors, frames, and ceilings buckled or fell inward. Despite the infrastructural damage, only 29 people reported receiving cuts or bruises. A smaller number, including those propelled into the air, experienced concussions. Remarkably, most of the 1,113 surveyed people who were outside during the explosion felt the heat of it on their skin—and 25 of them experienced some form of sunburn, including one individual who had been nearly twenty miles away. In total, 1,613 people asked for medical assistance. Of them, 112 were hospitalized, with the most serious harm befalling two: a child with a lacerated eyeball, and a middle-aged woman experiencing a spinal fracture. Thousands of people, for days and weeks afterward, remained deeply shocked and stressed.[37]

"It was dramatic. I did not expect to see models realized," said Popova, referring to simulations of the damage caused by asteroids—specifically those that do not reach the ground. That's right: even if an asteroid or comet fails to touch grass, it can wreak havoc. But how?

Imagine an asteroid plunging toward you. What happens? "It lights up," Mark Boslough, a physicist and explosion aficionado from the University of New Mexico, told me. That's because the space rock is battling with the atmosphere. At first, it's like punching through tissue paper. But the atmosphere gets denser and gloopier closer in, and the force against the asteroid rises exponentially. All that air accumulating at the front of the rocky projectile gets squashed, becomes really hot, and emits light. Unlike spacecraft, which have shields designed to direct all that nastiness around and away from it, an asteroid's Earth-facing side begins to cook. The pressure and temperatures continue to build, and parts of the asteroid begin to break off and tumble back into a thickening trail of glowing debris. The object fragments, the push-back from the atmosphere begins to slow it down, and all the energy of that hypersonic entity gets dumped into the atmosphere.

"At some point, it hits the point of no return," Boslough explained. Air injects itself into cracks before heating up, expanding, and forcing parts of the asteroid apart. Smaller and smaller asteroid shards are created in the fracas, all of which decelerate, donating even more energy to the sky. Suddenly, what moments ago was one single asteroid is now ethereal. "One minute it's solid rock; the next minute it's a big, hot cloud of gas," he said. "And it happens really, really fast," in a matter of seconds to fractions of a second. "It comes in, and then boom!" he gleefully exclaimed.

The explosion, known as an airburst, is a sort of hybrid between a sonic boom from a fighter jet and blowing up a lot of dynamite in a hot air balloon. The most energetic part of the airburst unleashes the most powerful shockwave, a wall of compressed air that radiates out in all directions and hits objects like a sledgehammer. But smaller airbursts leading up to the main event also produce tangible shockwaves. And unlike a bomb, these explosions travel along huge geographic areas as the asteroid keeps ploughing through the sky during its death dive. Minus radiation, "it's like a moving nuclear explosion," said Boslough.

So that's what happened at Chelyabinsk? Affirmative, he said. And that moving nuclear explosion, equivalent to 470,000 tons of TNT,[38] which precipitated all that damage and all those injuries, was caused by an asteroid just sixty feet wide, no longer than an average bowling lane.

I know what you're thinking. Sixty feet wide? You were expecting an asteroid at least the size of a famous landmark or skyscraper, right? Nope. Even a puny asteroid is enough to trash a city.

"We got, in a way, lucky," Jenniskens told me. The Chelyabinsk meteor disintegrated high in the sky, meaning its explosive energy was largely absorbed by the atmosphere. And it came in not directly perpendicular to the ground, but at a shallow angle; the blast waves that made it to terra firma were spread out over an expansive area. If the meteor had come in at 90 degrees, right atop the city, it would have penetrated deeper, and the force of the airburst would have been directed mostly downward. "Then the damage could have been higher," said Jenniskens. "The damage right below the entry trajectory would have been bad. People could have been killed, easily."

It seems patently absurd that such a nothingburger of an asteroid almost killed people. But, just like Boslough, Michael Aftosmis wasn't at all surprised. He also studies the physics of exploding asteroids—in this case, for NASA—and he eloquently boiled down why big booms have small beginnings. Dozens of things factor into the potential destructiveness of an asteroid airburst or ground impact, from the atmospheric drag on the asteroid to the angle it hits that atmosphere. But one variable reigns above all else: mass.

When an object moves, it has kinetic energy. And that can be represented by (don't panic!) an equation: $\frac{1}{2}mv^2$. That is: the speed of the object (v) multiplied by itself, then multiplied by the mass (m), then multiplied by 0.5. The amount of kinetic energy an asteroid possesses is a good approximation of how much energy the explosion will have. In other words, Aftosmis told me, this equation is a good way to guess "how big of an event this is going to be."

Seeing as the asteroid's speed gets multiplied by itself in that equation, you would think that that's the most important factor in determining an airburst's hypothetical ferocity. And speed *is* important, considering these things come in at tens of thousands of miles per hour. If they're fast, they're furious. But the mass of the asteroid is the key to the potential for violence here. The mass of something—kilograms, ounces, imperial tons, whatever you'd like—depends on a few different things. But crucially for planetary defenders, mass also depends on the object's radius—the distance from the center of the object to its edge—multiplied by itself, then multiplied by itself *again*.

When it comes to asteroids, size matters most. An asteroid with a bigger radius has more mass. Even small increases in size equal major increases in explosive potential. If, for example, you double the radius of an asteroid, it gains eight times more kinetic energy. Aftosmis conceded that the consequences of this mathematical tweaking may be difficult to visualize, so he proposed taking the 60-foot Chelyabinsk meteor and transforming it, making it 120 feet long—two bowling lanes end to end. The airburst now becomes eight times more energetic, or ten times if you round up. "So rather than 500,000 tons, we're talking about 5 megatons," he said. That's 5 million tons of TNT, enough to inflict significant trauma and death on a directly impacted city.

It's also difficult to visualize 5 million tons of high explosives. But it doesn't matter; you don't have to rely on your imagination here. Nobody does. Because in 1908, Russia was hit by an asteroid just over twice the size of the one at Chelyabinsk. Forget broken windows and collapsed roofs: an area the size of Washington, DC, vanished in a barbarous instant.

In remotest Siberia, as a crisp, cloudless twilight bathed the forest floor, many Evenkis were asleep in their camps. These Indigenous nomads lived largely isolated lives, moving from place to frigid place with their small families and herds of reindeer, constructing fur-insulated tents called *chum* (pronounced "choom") near places where fires could keep them warm, animals could be hunted, and fish could be snatched from their rivers.

It was close to 7 a.m. local time on June 30, 1908. Akulina, her husband Ivan, and an old man named Vasily were in their chum. Sunlight was starting to drip toward them. A summer's day awaited, like any other. The wind was perambulating peacefully around them—until, in an instant, it wasn't. A light breeze became a pandemonic bellow. Akulina and Vasily were knocked forward as their chum crumpled. Through some preternatural force, Ivan's body left the ground, rose into the air, and was thrown against a tree 130 feet away, breaking his arm so severely that a bone protruded through a blood-soaked shirt.

Not far from Akulina, other families were woken by artillery shell-like thumping. The sky turned bright white and red before that same diabolical shriek assaulted them, knocking them and their chums over, like leaves being swept away by a hurricane.

A heartbeat later, Vanavara—a collection of permanently inhabited huts about twenty-five miles south of Akulina's chum—was hit. There, a Russian man named Semyen had been sitting on his porch, greeting the morning. Suddenly, "a fiery ignition was formed. The fire made it impossible to sit from the heat, my shirt almost lit up on me," he recounted. He closed his eyes, the light vanished—and an explosion pummeled him, throwing him from the porch so aggressively that he passed out. He quickly came to, heard a thousand thunderstorms rumbling around him, and found that the doors and windows had bent inward, and were laying in pieces all around him. In the distance, high in the sky, blue hues were replaced by coruscating reds, which soon drifted westward, glowing as they went.

Nobody knew what had happened, or how to react. Some found themselves rooted to the spot in terror. One Evenki family's father, who was outside during the event, ran back to his family, screaming that he saw "it," while swinging about uncontrollably. Shocked beyond description, he collapsed onto his bed and died almost immediately afterward. Nearby, the incinerated remains of deer could be found, alongside burned chums whose food stores had been destroyed and cooking utensils had been warped and melted.

Others, like Akulina's group, headed to a nearby river, struggling through confusion and injuries. And as they did so, the number of trees still standing dropped off substantially until no tree was left standing at all. "All around, we saw a miracle, a terrible miracle. The forest was not ours. I've never seen such a forest," she recalled. Fires danced among the ashen cauldron, covering the refugees with soot and stinging their eyes. With nowhere to go, and nobody available to help them, Ivan succumbed to his injuries soon after.

Those eyewitness reports, recorded by ethnographers, newspaper stories, and various scientific expeditions in the subsequent decades, were translated and bundled together in a 2019 study by Jenniskens, Popova, and their colleagues.[39] For decades, there was no doubt about it: this event—dubbed Tunguska, after a nearby river—had an extraterrestrial origin. It's often debated whether it was a comet or an asteroid (most scientists suspect the latter), but what is unquestionable is how lucky the world was that this location, of all places, was where the event took place. In a matter of seconds, 80 million trees in an 800-square-mile swath of forest—the size of many modern towns or cities—was either vaporized or flattened, twisted beyond all recognition.[40] What once was, now wasn't. Damage was reported up to 250 miles away from the explosion—a blast equivalent to 12 million tons of TNT. And yet, partly thanks to their nomadic nature, barely any Evenki families were in the area at the time. Only three people are known to have perished.

Like the rock that blew up over Chelyabinsk, this asteroid excavated no impact crater. It failed to make it to the ground, splitting into fragments in the sky. It was moving faster than its 2013 comrade—a jaw-dropping 60,000 miles per hour, or eighty times the speed of sound—but, as ever, what really drove the destruction was its size. Chelyabinsk's asteroid was 60 feet long. Tunguska's was about 160 feet across. That's just half the height of the Statue of Liberty: all you need to destroy a city and kill millions. And it almost did. Had the Tunguska asteroid plunged toward Earth just three hours later, it would have hit Moscow with the force of 1,000 atomic bombs, and the world as we know it today would never have come to pass.[41]

Don't get me wrong. Most asteroids are affable, not people-killing assholes. Take Olivier Hainaut's word for it: he's an astronomer at the European Southern Observatory, which is run by a collection of nations whose telescopes can be found in the stargazing nirvana of Chile's Atacama Desert. They often take a gander at asteroids. "People don't realize that most—like, 99.999 percent of asteroids—will never come close and are absolutely harmless," the insouciant Hainaut reassured me. And even those asteroids that do head our way, through gravitational disturbances or the occasional collision, are eaten up by our atmosphere. "Most of them are pebbles and grains. We see them all the time as shooting stars."

But as Chelyabinsk, Tunguska, and Chicxulub underscore, with varying degrees of lethality, some asteroids *are* assholes. Hainaut ranks them by their size. Thirty feet across? "Annoying," he said. That could damage a city. Three hundred feet and bigger? "Dangerous." That would seriously wreck, if not outright destroy, a city. Three thousand feet and up? "Horrible." That could wipe out a large country, or much of a continent. Bigger than that, and you've got yourself a civilization-traumatizing to humanity-ending catastrophe.

These asteroids don't hit us with equal frequency. Pop culture and blockbuster movies suggest it's the bigger beasts, the 3,000-foot-and-up asteroids, that you should worry about. But giant asteroids are not only far less common, they are also the easiest to find, and astronomers think they've found almost all of those wandering close to Earth. It's those smaller asteroids, those city killers and country crushers, that are the real issue here: there are so, so many of them, and most of them haven't been found. And it is only a matter of time before some of them find us.

Planetary defenders are concerned with near-Earth objects, or NEOs, 99 percent of which are asteroids.[42] Astronomers refer to the huge distance between Earth and the Sun as one astronomical unit. An asteroid (or comet) becomes an NEO if it is 1.3 astronomical units away from the Sun. As of mid-2023, astronomers have found 32,412 near-Earth asteroids,[43] from the behemoths with planet killer potential to those that could reproduce Tunguska. And despite the ominous-sounding term, most NEOs are not a danger to us for the next one hundred years or so. Their circumsolar journeys are known with remarkable precision because the laws of orbital dynamics are unchanging and beyond reproach. The math is solid. Astronomers are awfully good at their jobs.

Skywatchers have discovered 850 NEOs of the 3,000-foot-and-above planet killer variety. Projections suggest that around 50 or so are yet to be found—not ideal, but far from a terrifyingly huge margin. Being so rare, these asteroids strike Earth approximately once every 700,000 years.[44] That's often enough for astronomers to want to find the remaining 50, but not so often that scientists are in a constant state of nihilistic dread.

An event like Tunguska happens once every 1,000 years. That isn't a hugely helpful number: not only is it a very rough estimate, but it also doesn't mean that the next one is coming along in the year 2908. One could land on your head before you finish this chapter, because out of about 230,000 Tunguska-size NEOs thought to exist, only 7 percent[45] have been found. Trying to work out how often they hit Earth is a sketchy enterprise.

What causes the most concern, though, are asteroids 460 feet across—the same size as the one I heartlessly threw at Seattle in the prologue. When I asked the cheerful Michael Aftosmis, our exploding asteroid specialist at NASA, about these asteroids, he became noticeably muted. "We would be talking about something that was over a thousand times more energy than Chelyabinsk," he told me—500 *million* tons of TNT. That's an order of magnitude higher than the largest nuclear weapon that has ever been tested above ground. Asteroids of any kind don't have a preferred place to impact, which means if one did hit Earth, it would most likely impact the ocean. One of these modestly sized asteroids plopping into the middle of the ocean may cause a significant splash, but nothing close to a country-swamping tsunami. Anything near a coastline, though, could easily drown nearby towns and cities. Any city taking a direct hit would be destroyed, and any impact within any densely populated state or country would be a mass-casualty event. Or, as Hainaut put it: "They are not going to cause a billion deaths, but they could still cause a million deaths, which is not great."

Here's the crux of the problem: Earth is hit by one of these asteroids every 20,000 years or so. And nobody knows when that next impact will arrive. There are 25,000 of these NEOs, and about 15,000 of them are yet to be found.[46]

Your lifetime risk of a city-slaughtering asteroid impact is low—but not vanishingly low. It's something that could happen during your lifetime, or during your children's lifetime. Any country could be hit. Any city could be destroyed. "The Earth is going to get impacted again in the future. It's just a fact of life in the solar system," Lindley Johnson told me, almost sedately. Why take the risk and do nothing about it? Why leave it to chance? Why condemn millions of humans, whomever they may be, to an avoidable death?

By the time Chelyabinsk rung a planetwide alarm bell in 2013—with the Chicxulub discovery, Shoemaker-Levy 9's impact, and the specter of Tunguska swirling around in the ether—NASA agreed that something needed to be done to defend the planet. Even the infamously dysfunctional US Congress had chimed in by this point: in 1998, spurred on by witnessing Jupiter's meet-cute with a comet, they directed NASA to find 90 percent of those 3,000-foot potential planet killers within ten years. (It took twelve years, but they did it.) And in 2005, Congress told NASA that, by 2020, 90 percent of those 460-foot city killers needed to be found.[47] (Whoops. Not quite there yet.) But it took the shock of Chelyabinsk—a destructive asteroid not spotted by any telescopic surveys—for NASA to agree that Johnson's 1994 wish to establish a world-saving scientific command center was, in fact, worth granting.

Johnson joined NASA in 2003. But it took until 2016 for his dream—

and, by that point, many of his colleagues' dreams—to be fulfilled. That year, NASA's headquarters inaugurated the Planetary Defense Coordination Office,[48] and made Johnson the world's first planetary defense officer. His job, and that of his staff, is to find those remaining NEOs, work out which ones are hazardous, and—no pressure—find a way to stop them from hitting the planet. He's undoubtedly happy with how things have turned out. "How many people get to create their job?" he said, chuckling.

Kelly Fast, the manager of the Near-Earth Object Observations Program, is essentially Johnson's second-in-command: she liaises with the government and the media, while keeping watch over the office's portfolio of planet-saving projects. "I hold the umbrella to keep the bureaucracy from falling on the people that are actually doing the work," she told me—the work to turn the planet-saving, childhood-era ideals of *Star Wars* and *Star Trek* into a reality. "It's about finding asteroids before they find us," she said. "And getting them before they get us."

How, exactly, are we supposed to do that? Scientists and engineers had asked themselves the same question since Gene Shoemaker got a few people together to talk about it in Colorado more than forty years ago, with many others contemplating the same conundrum for many decades prior. How would you go about saving the future from such a nemesis? How would you even practice killing an asteroid's chances of falling to Earth?

Then, one frosty day in 2011, someone came up with the most brilliant idea.

II

This Is for the Dinosaurs

Andy Cheng is the chief scientist for planetary defense at the Johns Hopkins University Applied Physics Laboratory in Laurel, Maryland—an hour's drive north of Washington, DC. Applied physics is a contemporary form of magic, the study of how you can turn the language of the cosmos into technological sorcery. That includes defending the planet, and it's just as well Cheng is the university's lead on that monumental undertaking, because, two years before the Chelyabinsk meteor lit a fire under NASA's feet, he had a lightbulb moment.

"It was a winter morning in early 2011," he recounted to me and a huddled mass of reporters on a visit to the laboratory's campus in the summer of 2022. While in the middle of exercising, "I had in my mind that we should really do a planetary defense mission. It's time, we should do one. A really good one." That's some top tier daydreaming right there, I thought. The soft-spoken Cheng had in mind something called a kinetic impactor—which means building a spacecraft, pointing it at an asteroid, and crashing into it, with the hope of deflecting it away from wherever it was originally heading.

That's it. That's all it takes to stop an asteroid heading toward Earth.

Alright, you got me—that's not all it takes. There are many facets to planetary defense, and aggressively prodding an asteroid is not the only method we could deploy. But you have to start somewhere, and that is with deflection: changing an asteroid's trajectory around the Sun by, well, hitting it with something. Space billiards, if you will. Nudge an asteroid onto a different orbital racetrack, and you've demonstrated that this technique works—and could one day be used to save the planet.

The straightforward-sounding prospect of deflecting an asteroid was not, however, Cheng's idea; it had been discussed among planetary defense philosophers for decades.

In 2002, a group of European academics, led by mathematician Andrea Milani, reasoned that you could send two spacecraft to an asteroid: one to impact it while a second would record the blast and study the aftermath. You couldn't just send an impactor by itself, because the moment it hit its rocky target, the signal would cut out, and you wouldn't be able to see what happened. Telescopes on Earth might be able to spot the spaceship-asteroid confetti scatter into space, but unless you sent an unimaginably massive spacecraft into the void to meet its maker, the orbit of that asteroid may only undergo a minimal change, making it impossible to see if the mission succeeded from tens of millions of miles away. So, they thought, the impactor spacecraft needed an ally.

The hypothetical mission was named Don Quijote,[1] a reference to the story of a low-ranked noble and his squire. The European Space Agency loved the idea, and reckoned it would be a great way to test the viability of this planetary defense technique. The problem with the idea of two spacecraft, though—especially in the years before the notion of applied planetary defense became a formal and properly funded endeavor, both in America and in Europe—is that it would cost a lot of money. And with not all the European Space Agency member states willing to pay up, this thought experiment remained just that: a thought, a fun thing to bring up at scientific conferences before moving on to other things.

Then, in 2011, Cheng had his epiphany. We don't need two spacecraft. We need two asteroids: one big one, and a smaller one, orbiting around its larger sibling in much same the way the Moon revolves around Earth. This asteroid pairing is known as a binary. Seen from the top-down, a binary would look a bit like a clock: one asteroid, in the center, with a smaller asteroid satellite going around it, like the tip of a clock's second hand. Or, if the binary system is seen from the side from Earth's perspective, it would appear like a lighthouse. Both objects reflect some sunlight, making them shine a little. When the moon-like asteroid has swung behind the bigger one, the binary's total brightness dips; when it comes back around so we can see both the moonlet and the bigger asteroid, more sunlight is reflected, making the binary appear brighter.

"We can target a binary asteroid," said Cheng. "You change the orbit of one of the two asteroids around the other one, and you can measure that change without a second spacecraft. You can do it from Earth." And, crucially, shifting the orbit of the asteroid moonlet would not put Earth at risk. Knocking the larger one could inadvertently push the binary into a future collision course with our planet. Prodding the smaller one would measurably alter its orbit around the bigger one, but it would not change the pair's orbit around the Sun in any meaningful way.

Cheng took the idea to NASA—and they considered it so elegant that they gathered some cash, teamed up with the Applied Physics Laboratory, and decided to make it happen. Earth was going to conduct a test like none other, a technique that may one day stop a genuine city killer or country crusher, perhaps even a planet killer. Cheng was, quite rightly, jubilant upon receiving the news, but not because of the legacy that a successful kinetic impactor test would have on human civilization. Like many of us, he just liked stumbling upon a really good idea. That chilly morning in 2011, he didn't think about how he'd just come up with a way to practice stopping a potentially killer asteroid. According to Cheng, all he thought at the time was: "I know how to do something!"

Conveniently, Cheng's hypothetical spacecraft already had a compelling target. Binary asteroids aren't particularly rare—scientists reckon about 16 percent of all asteroids we see are binaries[2]—but Earth couldn't target just any of them; it had to be one that was not too far from Earth, but in its own patch of observable, unobstructed sky easy to examine from the ground. It had to have a clearly mapped-out orbit around the Sun. And, crucially, the moonlet should be close to that murderous sweet spot of 460 feet long—the size of one of those plentiful and perilous asteroids that could, one day, find their way to us.

Perfection came in the form of a 2,600-foot NEO, dubbed 1996 GT, discovered by an Arizonan telescope that year. It seemed like a regular, lonely asteroid until 2003, when a close approach to Earth revealed its bijou companion,[3] a mere 520 feet long. The pair were baptized by the church of astronomy: the larger one was named Didymos (from the Greek meaning "twin"), and the moonlet was called Dimorphos (from a Greek word meaning "two forms"). Like clockwork, the latter orbited the former once every 11 hours 55 minutes. Neither posed a threat to Earth, but the moonlet was as ideal an impact candidate as you could find. And Cheng wanted to send a spacecraft millions of miles into space, crash into Dimorphos at 14,000 miles per hour, and get every single telescope on Earth and in the heavens above to watch, hoping to see that orbital period drop by just a few minutes.

The mission was given a name: the Double Asteroid Redirection Test, or DART. Johnson, Fast, and the Planetary Defense Coordination Office were on board. The Applied Physics Laboratory was irrepressibly eager to realize Cheng's vision. Now all they needed was a team to build the thing and shoot it into space.

This is that point in a heist movie when the big-picture thinker goes around hoping to recruit their ideal team of sleuths and thieves—or, in this case, pilots, programmers, astronomers, engineers, and machinists.

First off, though, let's meet DART itself—its final design iteration, after it went from a kernel of an idea to a fully formed blueprint that would impress any asteroid smasher. Imagine a van-size box, weighing about the same as an adult cow, with unfurling wings. That's DART.[4] At just 3.9 by 4.3 by 4.3 feet, it doesn't take up too much room when its lengthy solar arrays—each 28 feet long—are tucked inside. All the technological wizardry goes inside or onto that box.

DART has eyes, because even a soon-to-be-dead spacecraft cares about seeing where it's going. Its primary oculus is named DRACO, a wyvern-sounding name that stands for Didymos Reconnaissance and Asteroid Camera for Optical navigation. Like many fanciful monikers in science, it's a backronym: someone assembled the letters first to make a cool-sounding name, then filled in the corresponding words after. (No complaints here.) DRACO is essentially an extremely good camera lens. DART's secondary aperture, the star tracker, is exactly what it sounds like: as DART sails through the black ocean, DRACO focuses on Didymos while the star tracker makes sure all those tiny jewels are where they are supposed to be, practicing the same sort of navigation technique Earth's seafarers once used to find their way to shore.[5]

DART doesn't have any winds to catch in its sails, though. While the solar energy provided by its wings powers its internal circuitry, its thrusters burn hydrazine fuel to let it soar forth. Another booster, one not critical for the mission but certainly useful, is also attached to demonstrate its feasibility for future missions. This experimental device, NEXT-C (NASA's Evolutionary Xenon Thruster—Commercial) pushes electrically excited particles out DART's derriere, contributing to its intentionally doomed journey.

Although maneuvering commands will be sent from Earth, the spacecraft will be so far from home that any communiqués will take thirty seconds or so to reach it by the time it's close to its destination. Hitting the relatively small Dimorphos with the extremely small DART is a tall order for a human pilot trying to work out what's happening 7 million miles away, so instead, most of the flying will be done by the spacecraft itself. Its Small-body Maneuvering Autonomous Real Time Navigation, or SMART Nav, is effectively DART's brain.[6] Although constructed on a foundation of code used to guide ballistic missiles to their targets, SMART Nav has a bespoke collection of algorithms designed to target potentially killer asteroids, not people. During its terminal journey, SMART Nav will use what DRACO sees to make sure it's heading toward Didymos, as it grows from a fuzzy handful of pixels to a great big rock. At the last moment, it will spot Dimorphos emerging from behind Didymos's shadow, lock onto its target, and . . . boom.

This thrilling sequence of events all relied on DART being put together with painstaking precision by some of the world's best engineers. We should thank our lucky stars, then, that the Applied Physics Laboratory had Betsy Congdon in its employ. "I lead the team that's responsible for actually assembling everything. Like, physically putting everything together and making sure it's all in the right place is my job." In other words: Congdon and her cleanroom-besuited contingent of engineers and technicians were tasked with solving a jigsaw puzzle of labyrinthine circuitry, doodads, and thingamajigs. And as they did so, a voice in the back of their minds constantly harangued them, pointing out that if they made just one tiny mistake, the spacecraft might flounder, and the mission would fail.

To me, that sounded like an unhealthy amount of pressure. Congdon didn't come across as the type to thrive on that sort of pressure, necessarily. She just really loved making things, disentangling things, and getting to know puzzles inside out.

"I will say that this is the job I've wanted since I was a little kid. I didn't really know what this job was, but like, space . . . taking things apart, putting things together, thinking about how things work—in space!—is something that's always fascinated me." Thankfully deciding not to use her mechanical magic in service of an evil galactic empire, she started the preliminary design work on DART in 2017, when plans were "really coming together, like a real mission." In late 2018, the mission entered its critical design phase, the point at which the blueprints get finalized and you start gathering thousands of jigsaw pieces. Everything was going smoothly until they reached the integration and test phase, the part when the spacecraft's components begin to be glued together and rigorously tested—because that phase began in April 2020.

It's difficult to recall a more stressful time than April 2020, just a few weeks after the World Health Organization declared that a new, frightening pandemic was officially upon us. How on Earth did Congdon and company handle making DART in those early days of COVID-19?

"It . . . was a very interesting time," she said. Hmm. But isn't assembling a spacecraft intense enough as it is? A firm nod. So, add a lethal virus atop that, and . . . "It was like an intense time on steroids." But they just had to get on with it. DART couldn't be launched whenever they felt like it. All space missions have windows, small gaps of time in which the craft can fly into the beyond and reach its target before it shifts into another, more inconvenient corner of space. DART needed to fly in November 2021. It had to be built, COVID-19 be damned. "We spent a lot of time in the clean room, together, masks on, putting together a spacecraft, because that's what needed to be done," she said.

DART's limbs and organs didn't appear overnight at the Applied

Physics Laboratory. Although some viscera, like DRACO, were manufactured on-site, others were made by various companies or organizations: the propulsion system was made in Washington State; the rollout solar wings came from California; the NEXT-C ion engine was built at NASA's Glenn Research Center in Ohio. And when all of DART's components arrived in Maryland, where they were tested repeatedly to find the slightest hint of a fault, to make sure they could survive the wild temperature extremes of outer space, and carefully pieced together into one cohesive whole, there was only one thing left to do: simulate an earthquake.

I'm sorry, do what? "You literally shake it," said Congdon. For about a week. "That's always exciting." After all, it's ultimately going to be installed inside a giant rocket that will blast it into space, so you want to know it can handle tremendous tremors. In the months leading up to the launch in November 2021, everything seemed like it was coming together. "I'm just as excited as everyone else for impact day," she told me—everyone else, I'd argue, except for one team member: the one whose job it was to guide DART to the X on its benighted treasure map.

Elena Adams is DART's mission systems engineer. But you might as well call her the spacecraft's lead human copilot. Thanks to software like SMART Nav, DART's navigation through space would be partly autonomous, relying only on the lightest of human touches. And on its death dive toward Dimorphos, during the mission's last minutes and hours, its cybernetic brain would take over completely. But, aided by stellar navigators from NASA's Jet Propulsion Laboratory, Adams and her team would have to talk to DART and command it to make the small-but-significant adjustments necessary for it to get anywhere close to its target, all while making sure its mechanical vital signs looked great. Use up too much hydrazine fuel, and DART's done. Accidentally trigger a malfunction in an instrument, and the spacecraft could drift off into the darkness.

Adams's job is unique—not just on the mission team but in the history of spaceflight and interplanetary exploration. After overseeing the construction of the spacecraft, and after it was eventually launched into space in November 2021, she would have to make sure DART is kept alive right up until she has to make sure it died. This is not normal for space missions.

When scientists launch a spacecraft costing hundreds of millions or even billions of dollars, all they want is for it to be kept alive and well, doing all the science it can, for as long as possible. Deploying a spacecraft outside Earth's atmosphere is fraught enough as it is, but if it then must travel to and land on another planet or moon, the anxiety becomes acute. For instance, NASA even has a phrase for the moment a rover or lander attempts to make it through the Martian atmosphere on the way to the surface: "seven minutes of terror." Afterward, the rover's team hopes it can keep on conducting science for as long as the planet will let it. When this robot eventually dies, months or years later, usually through technical failure or environmental aggression, the team—particularly those who invested decades of their life working on the mission, from cradle to otherworldly grave—mourns its loss in a way not dissimilar to the way people mourn the passing of loved ones.

DART would be different. At the end of its ten-month voyage, the aim is to crash into Dimorphos and cause DART to die as spectacularly as possible. A glorious death is the entire point of this mission. "It has a very definite end," Adams told me. At that moment, at that exact second, the mission team wants to see its heartbeat flatline. "In this case, if you keep it going, you really messed up. You want it to die." Would Adams mourn its demise, like its Martian cousins? "No! I will not be sad. If it died like it should, then I will be extremely happy," she said, through an almost nefarious grin. "I'm not viewing it as the death of a spacecraft. I'm viewing it as the birth of a whole new planetary defense program to actually protect ourselves from asteroids. How cool is that?"

I didn't know what to expect from DART's human copilot, effectively the flight commander tasked with demonstrating that we could deflect a killer asteroid away from Earth. Those who have similar roles, in my experience, are often a little highly strung, understandably nervous about the mission they are about to shepherd. If Adams was nervous during any of our conversations between 2021 and 2022, she hid it well. She was full of verve, unfailingly loquacious, always overflowing with excitement at the thought of how unbelievably badass this mission was.

She was also cognizant that, unlike most space missions, DART didn't need someone translating jargon-riddled scientific speech into colloquial parlance. "I don't think there's anybody that hears about planetary defense and goes, 'You know what, we really shouldn't do that,'" she said. "Everyone knows the story of the dinosaurs. Pieces of rock falling from the sky affect you." But she acknowledged that the public can be a little disconnected from the reality of co-showrunning a space mission. To many, it doesn't feel like it could be a real job. "People usually react pretty strongly, mostly with: 'They pay you for this?' And I'm like, yeah!"

Like many of her friends, she grew up curious about the stars. "I went to an astronomy camp as a kid," she told me. The telescopes, the sights, the twilight hours—being a nocturnal sky watcher strongly appealed. But the solar system had a more compelling draw than those ineffably distant galaxies, and after a stint studying the hazy Saturnian moon of Titan, she decided that even those weird worlds were a little too far from home. She wanted to visit them—if not in person, then with robotic envoys. At that stage, "the science part involved me sitting on the side of a really dark lab writing spaghetti code, which I... I don't know. It was too much for me. I thought it was much more exciting to go out there and build things. And I love working in teams." She did the video game equivalent of going back to the character selection screen and changing her stats: she got another degree, this time in spacecraft engineering, and ended up at the Applied Physics Laboratory.

Adams has worked on several solar system missions, from spacecraft that preferred the white-hot inferno of the Sun to those preferring the chillier climate of the outer solar system. But DART is her magnum opus, not a science mission but a technology demonstration, one in which the engineers, not the scientists per se, are front and center on the stage. "That's what makes it more exciting," she said. And the fact that DART has "a noble goal everyone can get behind."

Like the rest of the team, she rippled with impatience in the months and weeks leading up to the November launch. Space was unfeeling, nonconscious, but had occasionally thrown rocks and ice balls to get us. "And now, you're going to get space," she said. If, that is, DART worked as it was supposed to. During our conversation, she picked up a handheld, cube-shaped model of DART and held it above her desk. "Very nice," she said—before a piece ominously fell off it and landed on her desk. "A piece of it just fell off on me... not like in real life," she clarified.

It turned out that she did get nervous, when she really thought about it, joking that she needed more dye in the past year or so to cover up gray hairs. But channeling those nerves into golden retriever energy wasn't just an act of self-care. She led a team of engineers, all taking emotional cues from their leader. Adams hoped to keep them from getting psychologically contaminated as best as she could. "As a person, I'm pretty excitable. And a healthy paranoia is part of my job."

DART cannot inaugurate Earth's planetary defense system by itself. Earth's astronomers have a starring, and difficult, role to play, too. Not only must they track a fast-moving explosion 7 million miles from home, but they also must study the binary asteroid system long before, and for a few months after, DART's death day. Just as you cannot tune in to a movie at the climactic scene and hope to fully comprehend the entire plot, you cannot say how well the deflection campaign went if you only see the explosion—the hero's journey, and the epilogue, are just as vital to the endeavor.

This wouldn't be a case of astronomers just doing what they thought may be best, observing the asteroid pair when they had a spare moment. These watchers needed someone standing on the wall, making sure everyone was pointing everything at the right patch of sky at exactly the right time. Someone had to commandeer as many telescopes on Earth as possible to record this once-in-a-lifetime experiment as it played out. All actors need a director. And DART needed Cristina Thomas.

Thomas, a genial planetary astronomer from Flagstaff's Northern Arizona University, grew up in Chino, California—basically a dairy farm at the time, she told me. In such a quiet setting, her career choice could have been anyone's guess. But she was, like many of her peers, swayed by the power of two things. The first was America's increasing investment in space exploration. Thomas glimpsed the first images from the Voyager program, a pair of space probes that gave humanity its first up-close-and-personal look at Jupiter and Saturn, when she was in elementary school. "They put those pictures on stamps; they were everywhere." In 1997, while she was in high school, NASA landed the Pathfinder robot on Mars. She also happened to live down the road from the Jet Propulsion Laboratory; when she visited, scientists were giving out space-based paraphernalia. "It kinda felt like a rock star was handing me a poster." And that second overwhelmingly strong influence? A visually arresting, Wookiee-filled space opera, of course. "I was also a gigantic nerd," she said, matter-of-factly. "I was really into *Star Wars* and all those other things. I was really primed for this."

Like Adams, she also differentiated astronomy from planetary science, speculating at university that studying worlds humans can visit (one way or the other) would be supremely gratifying. "Galaxies, not so much," she said. She gestured toward an academic tome on her bookshelf, one telling tales of meteorites and their origins. A figure in that book revealed that scientists can look at the chemistry of a meteorite, compare it to an asteroid out in the nether, and reconstruct that asteroid's journey from space to Earth. "I was completely mind blown that you could do something like that."

What blew *my* mind was that she had developed an ambitious scheme to politely seize control of every capable observatory on Earth and use them to see DART's all-important collision from every possible angle. She had begun to recruit telescopes and their astronomers from New Zealand, Australia, South Africa, Israel, Kenya, Spain's Canary Islands, Italy, West Virginia, New Mexico, California, Arizona, Texas, Hawai'i, Chile, and Argentina[7]—and had schemed to commandeer even more by impact day. From the middle of deserts to the tops of leaf-veiled, glacier-carved hills, hundreds of astronomers around the globe would do their best to line up their scopes, hoping not just to capture debris flying off Dimorphos, but to chronicle a change in its orbit around Didymos.

Thomas's worldwide campaign wasn't confined to Earth. Lucy, a spacecraft on its way to visit—but not crash into—a family of asteroids, would turn to face Dimorphos on impact day. The venerable Hubble Space Telescope would join the party, too. So would the James Webb Space Telescope. Sitting about a million miles away from Earth in a gravitationally stable dip in the fabric of space-time, this $10 billion observatory[8]—the most expensive single spacecraft in human history—was built to find galaxies and stars born just after the big bang and help us unravel the origins of the universe as we see it today. But Thomas had convinced Webb's team to lend it to her, to capture the moment Earth proved it can defend itself from asteroid strikes.

The scale of such an operation, involving the synchronous actions of high-tech machinery all over—and beyond—the world, is normally the sort of thing you would expect of a supervillain. I suggested this to the anything-but-villainous Thomas. "It *is* kind of satisfying," she said, smiling.

Her populous team of Dimorphos watchers were no less fervent about the operation. But conversations revealed a few cultural differences in how the experiment was being perceived. Most in the United States talked about DART's meeting with Dimorphos using explosive terminology. Unlike many of her peers, Michele Bannister, a planetary astronomer at the University of Canterbury in New Zealand, did not think of the mission as something quite so uniquely percussive. Impacts between celestial objects happen all the time. "This is how planetary bodies have conversations," she said, and DART "is us having a conversation with an asteroid." Thanks to Thomas's plotting, the world would get to eavesdrop.

At the start of the DART project, it was easy to list the many things that could go wrong. The brain of the spacecraft may break down. Its thrusters might not work. DRACO may go blind. DART may hit the target, and nothing of any consequence may happen. It may miss the target. Hell, the thing might blow up on the launchpad.

All hypotheticals for now. What mattered above all else was that America, home of the brave, was going to give it a shot. They were leading the charge. And Europe didn't want to be like all the other countries in those 1990s Hollywood blockbusters, huddled inside German castles listening to the dramatic proceedings unfold over the wireless, or watching TV while wearing berets and wielding baguettes in Parisian cafés. They wanted to get in on the action, too. Much like Cheng seeing the potential of the asteroid binary Didymos and Dimorphos, the European Space Agency also thought two was the magic number. With DART starting to go from cool concept to rock-solid reality, Europe—taking a page from the Avengers playbook—suggested a team-up, wherein DART would be accompanied by a second observational spacecraft after all, one funded by European states. The Asteroid Impact and Deflection Assessment, or AIDA initiative, was the best of all worlds: the impact would be seen by astronomers up close through the spacecraft's eyes, and from afar via Earth's own telescopes. And what would Europe's spacecraft be called? AIM, of course: the Asteroid Impact Mission.

But like Don Quijote, AIM died at the hands of the bean counters. In 2016, at a gathering of member states, AIM only reached 70 percent of its funding, and fell behind as DART's squad of superheroes began to crystallize.

Patrick Michel, speaking with a musical French accent and through a waterfall of ennui, sighed at the memory of AIM's abortive saga. "I was born in Saint-Tropez, the famous village in the South of France. My parents had a five-star hotel, and I was growing up with show-business people," he told me. "But stars of showbusiness are less easy to understand than the real stars. Less rational." While at school, he came to realize that mathematics and physics could be used to translate what appeared to be messy and manic aspects of existence into quantifiable and predictable stories. And if you turned that power to space, he said, "you can predict things, and you can understand how stars work without touching them." His future in planetary defense was all but guaranteed when he studied NEOs for his PhD thesis, and by 2016, he was one of AIM's gleeful zealots.

So, what happened? "Germany," Michel said, crumpling his brow. Often one of the biggest financial contributors to the efforts of the European Space Agency's space exploration efforts, they were expected to support AIM's conception. But during the December 2016 negotiations held in Switzerland, they dropped out.[9] "They withdrew at the last moment," giving the director-general of the European Space Agency no choice but to scuttle the nascent mission. Michel did little to hide his frustration when I asked why Germany tanked the project. "Mm, don't ask. Don't ask. It's politics," he said. These gatherings, which happen every three years, are like open-air markets filled with rancorous vendors, all hoping to get what they want for the lowest cost. Suspecting something was also going on behind the scenes, but not willing to speculate or get into specifics, he simply threw his arms up in the air.

Michel, hoping to exhume and reanimate some of AIM's remains, went to see the director-general. "I tried to convince him it was worth finding a solution," he said, with a slightly cartoonish tone of menace. He succeeded. Hera—named after the Greek marriage deity—was born,[10] a slightly streamlined and watered-down version of AIM. Equipped with cameras, hyperspectral imagers, thermal detectors, laser altimeters, and all kinds of Dimorphos-surveying high-tech gadgetry—along with two petite robotic satellites of its own—DART's companion was given European assent by member states at the 2019 gathering, and Michel was christened as its commanding officer.

But DART would be arriving at Dimorphos in the fall of 2022. Even adhering to the quickest of construction and launch schedules, Hera wouldn't get there until the end of 2026. It would get to see what post-impact Dimorphos was like, a valuable bit of scientific work. But it would be extremely late to the party, arriving long after DART's death. All that would be left was America's trash to marvel and tut over.

During this saga, Italy ran out of patience—specifically, Simone Pirrotta and members of the Italian Space Agency. They decided they were going to make their own kind of music. In 2017, after AIM was pushed unceremoniously off a cliff, Pirrotta found himself at the Italian embassy in Washington, DC, having been invited to give a talk and to strengthen the connective tissue between his space agency and NASA. Pirrotta spoke about CubeSats,[11] a type of small satellite (sometimes as teeny as a sugar cube, but often as big as a toaster) that would later hitch a ride with the Hera mission. Although not as capable as their more sizable sisters, CubeSats are very cheap and extremely light, meaning at little cost they can be propelled into space to carry out several scientific activities.

After the meeting, people gathered for cocktails, and Americans and Europeans alike poured one out for AIM. Pirrotta, meanwhile, was plotting with a member of the DART team, and the two of them had conjured up a rather brilliant idea: Why don't we find some space inside DART for a CubeSat, built by the Italian Space Agency, and give it a camera? That way, it can be set free just before DART perishes and record the carnage up close. "This idea seemed very attractive," Pirrotta told me. There was only one condition from the American side. "We had to promise not to damage DART at all," he said, which seemed fair enough.

And with that, Pirrotta's superiors gave the thirty-pound, bread loaf–sized Light Italian CubeSat for Imaging of Asteroids, or LICIACube,[12] their benediction. It would fit snugly into DART's casing, and a couple of weeks before impact, a little hatch would open, and a spring-loaded mechanism would launch it into space like an overeager jack-in-the-box. Flying by Dimorphos just moments after it was wounded, it would use its two cameras—the narrowly focused LICIACube Explorer Imaging for Asteroid, and the wide angle LICIACube Unit Key Explorer—to take those history-making shots. That's right: the two cameras were named LEIA and LUKE. That is utterly delicious all by itself, but the cherry atop the mission's pop culture cake came from the DART team scientist who met with Pirrotta on that all-important DC evening in 2017: he made personalized Luke and Leia emojis for the DART team's private messaging channel. And that tells you almost everything you need to know about the profoundly cordial and professionally geeky Andy Rivkin.

Rivkin, a musically gifted, bristly bearded polymath, likes the occasional dive into history. In the late nineteenth century, "they would organize train wrecks, like actual train wrecks. Just because . . . well, we've got these trains, I don't know," he told me. Take the "Crash at Crush,"[13] for example, one such smash-'em-up that transpired not far from Waco, Texas, back in 1896. For some esoteric reason, it was determined that the best way to give the Missouri-Kansas-Texas Railroad some publicity was to get two locomotives and crash them into one another. About 50,000 people came to watch the crash, which was a little bit too spectacular: the explosion of the trains' boilers propelled a huge driving wheel into the sky before it smashed into the crowd, killing three.

Aside from the fatalities, DART is a distant spiritual successor to these organized train wrecks. This may be the simplest kind of test—seeing if humanity can move something in space by hitting it quite hard—but you've got to start somewhere. And everyone, including Rivkin, was hoping to get as much science out of the extravaganza as possible. "We are crashing our spacecraft into this asteroid—not for sport, but to do an experiment. And part of the experiment is measuring what you've done," said Rivkin. The two Andys—Cheng and Rivkin—are the DART investigation team leads. That means they need to extract as much scientific information from the impact as possible, and above all else determine if the deflection worked, and why it worked. "My personal job is organizing all the other scientists." Herding a thousand curious cats, he meant.

"When I was a kid, and I was in college and stuff, I had friends that were going to med school and doing this and doing that, and I was like, oh man, I don't wanna do anything that has too much responsibility," he told me, a month before DART's scheduled launch.[14] "Astronomy seemed pretty safe." And now he's one of the leaders of a planetary defense mission. How's that unexpected responsibility sitting with him, I wondered? Just fine, it turned out. "We get to do this really striking, almost sci-fi kind of thing. And it's not gratuitous. It's what we need to do."

Did he constantly think about the significance of this mission? Not especially, he said. "Mostly it's the 'this is cool,' honestly." But he's clear-eyed about the risks of not doing something like DART. He raised the oft-cited odds of a Dimorphos-size city killer impacting Earth, that once-in-20,000 years figure. Aside from the problem that those odds are little more than an educated guess, another issue is that, when you think about it, once-in-20,000 years is already discomforting. Let's round up a bit and say that people live for 100 years, and let's say that a city killer hits exactly once every 20,000 years, at random. That means there is a one-in-200 chance that this impact will happen during your lifetime.

Those odds are still relatively low. But "it's higher than you'd like," said Rivkin. "I think the numbers work out that, at the start of every decade, the universe deals us five cards. And if we've got a four-of-a-kind in those five cards, then there's going to be an impact that decade. You don't often get dealt a four-of-a-kind out of five. But it happens, you know?"

Wouldn't it be great to beat the house, to stack the odds in humanity's favor? That, to Rivkin, was what DART represented. It was the first test of our ability to defend the planet. It's not the only impact-preventing idea. "We want more tools in the toolbox. We want not just the hammer, but the screwdriver," he said.[15] But a kinetic impactor defense, like DART, was always going to be tested first, partly because it's not especially complicated (technologically speaking), and partly because, on paper at least, it seems like a very effective technique. And DART's success would pave the way for other tests, and eventually a world in which asteroids are just scientific and aesthetic marvels, not threats. "I think of it more as an insurance policy. If it's one less thing that anxious people have to worry about when they're trying to sleep, I think that's worth it—one less piece of existential dread."

His lighthearted nature and propensity for adorable sci-fi gifs made me suspect that, should DART hit Dimorphos, he'd have something cool to say as he celebrated. When I asked him about it, he referenced a blog post[16] he wrote back in 2018 about the mission titled "'Asteroids have been hitting the Earth for billions of years. In 2022, we hit back.' Some feel like that's a little over the top." When the mission was completed, perhaps he'd take inspiration from the more straitlaced mission members, those who emphasized how much this mission would matter to everyone's ongoing survival. Alternatively, "people have talked about this whimsically as revenge for the dinosaurs," he said. He'd see what felt right at the time.

DART managing to reach its target was one thing. But nobody knew much about Dimorphos at all, which meant nobody knew what would happen if the spacecraft hit it. The asteroid was simply so small and so distant that the first time astronomers would get a proper, detailed look at it would be through DRACO, mere minutes before the feed was severed. Dimorphos could be one big solid rock, or a pile of loosely bound rubble. It might not be round, but egg-shaped, look like a rubber duck, two bowling balls glued together, or even resemble a dog bone. "You hope it's something reasonable," Adams told me. One of her colleagues gave her a donut fridge magnet, a sort of talisman protecting against the asteroid having a big hole in the middle that a perfectly aligned DART would fly straight through.

Angela Stickle had thought about the multiverse of possibilities more than anyone else. No matter what Dimorphos was like, whether it had a deceptive pastry-like form or consisted of boulders flying in formation, and no matter how DART hit it, from a dead-on direct hit to a glancing blow, chances are she had virtually recreated it. Thanks to her team's cornucopia of computer simulations, whatever happened on impact day—except a miss, of course—could be compared with these digital dreams to help Rivkin's scientists understand what happened. Impacts are Stickle's jam. "I study things that run into other things really fast, and really hard," she told me. Asteroids are no strangers to such crashes.

I can relate to Stickle's passion. My own academic background, volcanology, was the result of two things: a lifelong love of *The Legend of Zelda* video games, whose splendorous overworlds always include fantastical volcanic realms, and the ineffable awe of seeing mountains explode. I asked Stickle, a distinctly convivial individual, if her fondness for impacts has similar roots. "I liked explosions when I was little, you know? Firecrackers . . . I used to bury them and watch little craters form." After she had studied aerospace engineering and geology, her future PhD advisor asked if she would like to do more of those childhood explosions, but this time for science. "Hey, NASA has a really big gun," he said at the time. "Do you want to come shoot things with it?" Only a fool would have said no.

Impact experiments, whether practical or computational, are usually deployed to work out how asteroids and comets cut pieces from planets and moons. Stickle was in the unusual position of leading a group that needed to work out what may happen when humanity punched an asteroid. All sorts of possibilities had to be considered in their simulations, and the simulations of their peers elsewhere. "The thing I'm most looking forward to is proving all of our models wrong," she said. "Any result is an interesting result. DART will teach us something no matter what happens." This sounded like the sort of thing that people would make bets on—what the asteroid is like, how much the asteroid's orbit around Didymos changed, how much Dimorphos gets pushed back by the impact, how much debris comes off it, and so on. Was it? A pause, then a small nod. "There is a pool," Stickle said—but she didn't elaborate on her own bets, or the potential winnings.

Win or lose, what mattered most to her was also what appealed to the mission's other team leaders: it was a seriously cool mission, one that happened to have incredibly high stakes. "We're purposefully crashing a spacecraft into an asteroid. It's one of those things that's literally out of a movie," she said. "And this type of technology could actually save millions of lives, which I don't think a lot of physical scientists get to say. You can't control space. But you can push back against it."

DART's mission team had been assembled, as had the spacecraft. In November 2021, it was loaded into a SpaceX Falcon 9 rocket at the Vandenberg Space Force Base in California. And when November 24 arrived, the weather looked clear enough to light the fuse. How was everyone feeling?

Adams was split between exuberance and melancholy. "You built this thing, and now you're just kind of saying goodbye," she told me. Not really to the spacecraft, though: her team of spacecraft medics and human copilots would be spending the next ten months making sure that DART fulfilled its fate. But many of the technicians and engineers she worked with for all those years prior—including during the worst of the pandemic—would be moving on. It was like trading in one set of friends, or a family, for another: both marvelous, but the best kind of farewells are never easy, nor should they be.

Congdon shared Adams's saudade. "Often, you're working on something you're never going to see again," she said. She talked about her previous mission, the Parker Solar Probe, a spacecraft already many millions of miles from home that, one day, will extinguish itself in the blazing light of our local star. "Most of the stuff, you watch it go off on a rocket, you close the rocket off, and you realize, in the moment that's happening, that this thing you've been spending years working on . . . you're never going to see it again." But that doesn't make each successive launch any easier on the soul. "Launch is a very emotional moment," she told me. But until DART met its fiery end, Congdon sensed that she would be "still kind of connected to it. It still feels like something that's real and going on."

Like several team members, Stickle couldn't travel to California because of COVID-19 concerns. She remained in Laurel, at the Applied Physics Laboratory, which put up a big screen on the night of the launch. "It was 1 a.m., and it was 15 degrees outside," she told me. "They had bonfires." Someone came up with the bright idea of screening an *Armageddon* and *Deep Impact* double bill. Having worked on DART since she started at the institute back in 2013, seeing it go from concept to sitting atop a huge rocket was bizarre. "It was a little bit surreal," she said. There was plenty of excitement. "And a little bit of terror, actually."

Thomas, like many others, simply could not believe DART was happening. "This had sounded like the coolest thing ever; we were super into it. But I never honestly thought, up until the very end, they would let us do this. I thought this was a very complicated thought experiment for many years."

Rivkin managed to get to California for the launch on November 24. The Sun had set; illuminated vapors were slinking off the rocket like phantoms. "At the launch, I was lucky enough that I was able to invite my brother; my spouse couldn't make it, but I was standing out with my brother, watching it go from outside of Vandenberg." As the final minute arrived, and that iconic countdown began, his brother started recording on his cell phone. Rivkin thought about doing the same, but resisted the temptation, instead choosing to be fully present in the moment.

Three. Two. One. Zero—ignition. That beautiful roar, that stunning explosion of light and fury. Humanity's very first attempt to protect the entire planet from a future cataclysm was on its way. The rocket sailed into the night, like a shooting star in reverse. Rivkin, often jaunty but not usually prone to paroxysms, couldn't help himself. "My brother posted the video to Facebook or whatever, and you can hear at the point where it's launching, I just start laughing maniacally, overcome with—oh my god, this thing's going!"

All being well, DART would impact and deflect Dimorphos on September 26, 2022. A lot had to go right over the next ten months. But the fact that this mission was even happening was itself astounding. The entire mission, its development, construction, launch, operation, and postmortem examination, cost $314 million.[17] That's about the price of a prodigious luxury yacht. Or, if you'd prefer, a rehearsal for saving the planet, a benefit to all seven billion of us, DART cost just 0.04 percent of America's $801 billion military budget for 2021.[18]

"This is humankind's first time that we have used our knowledge and technology to start to rearrange the solar system to something more to our liking," Johnson told me. We are making the solar system more habitable. "It's a milestone, threshold, whatever you want to call it, to the future." Did he have any plans to celebrate the day DART hopefully hit its mark? "If there's not champagne there, I hope they at least have a rum and Coke for me to celebrate," he said, with a hint of a smile.

Cheng, at the time of the launch, wasn't available to speak with me. I heard he was content to mostly sit back, say very little, and watch his planetary defender fly.

III

Hollywood and the Nuclear Hail Mary

I reckon it is safe to presume that, unless you are nihilist or a misanthrope, you would prefer it if Earth wasn't hit by a city killing or country crushing asteroid anytime soon. That's probably why you picked up a book titled *How to Kill an Asteroid.* But I can guess at another reason: such a saga, you suspected, is likely to feature some rather significant explosions—and big booms promise cinematic action. You will be happy to know that I cannot tell the story of planetary defense without expounding on the role of nuclear weapons.

It all started in 1998.

That year, a meteor shower came out of nowhere and blew up the *Atlantis* space shuttle in low-Earth orbit, killing its crew as they attempted to repair a satellite. After the hailstorm of space rocks turned New York City into Swiss cheese, a lone astronomer peered through his telescope and found an asteroid the size of Texas heading our way. The president of the United States got the lowdown from a senior NASA official: that gargantuan rock was heading to Earth. He asked what kind of damage to expect. "Total, sir," the official replied.

"It's what we call a global killer. The end of mankind. It doesn't matter where it hits; nothing would survive, not even bacteria." And then, with impeccable timing, a scientist burst into the closed-door meeting holding a piece of paper with something on it circled in red. "We have eighteen days," he said, breathlessly, "before it hits Earth."

Spoiler alert: it did not hit Earth. Bits did. A piece hit Shanghai, while another squashed Paris. Panic and death abounded. But the world was saved because a bespeckled scientist at NASA came up with an ingenious idea. That person—who the NASA official that briefed POTUS earlier described as "pretty much the smartest man on the planet"—immediately dismissed the idea of deflecting the asteroid. "You could fire every nuke you've got at her and she'd just smile at you and keep on coming," he sneered. Instead, NASA needed to get a nuke inside the asteroid itself and detonate, splitting the asteroid cleanly in half and sending both fragments away from Earth.

And how would NASA do that? "We drill," the NASA official said.

And it worked. In just a matter of days, the best oil rig workers in the world were sent into space with barely any astronaut training. They landed on that Texas-size titan, fended off the asteroid's natural defenses, just about succeeded in excavating an 800-feet-deep hole, dropped a nuclear warhead into it—which the US military tried to remotely and prematurely detonate—and pressed that all-important red button at the very last second. Armageddon was averted. To celebrate, Aerosmith sang "I Don't Want to Miss a Thing." And the citizens of the world gave NASA and nuclear weapons their collective blessing. The end.

I was ten years old in 1998, and it was a formative year for two reasons. The first was that *Ocarina of Time*—one of the greatest video games ever made—was released for the Nintendo 64. The second reason? Bruce Willis in the movie *Armageddon* saved Earth from a planet killer asteroid by drilling into it, burying a nuclear bomb, and blowing it up.

I think even at the time, as a child watching Michael Bay's bonkers blockbuster, I knew it was a bit silly. At one point, Steve Buscemi's character—who is suffering from the very serious and definitely real "space dementia"—sits on a nuclear bomb and rocks it back and forth, pretending to be Major Kong from *Dr. Strangelove*. But the many empty-headed moments aside, this popcorn flick about defending the planet not from supervillains or aliens, but a threat that genuinely exists, thrilled my young self to the point that my fascination with asteroid impacts never faded.

Unsurprisingly, *Armageddon* is as plausible as a psychic pineapple. The movie, and its cinematic cousins, get almost everything wrong about planetary defense—everything, that is, except the part about nuclear weapons. You wouldn't send oil rig drillers onto an asteroid. But in certain situations, it turns out that using a nuke may be not only Earth's best option, but the only thing that stands a chance at saving the world from calamity.

Megan Bruck Syal isn't quite sure how to tell people that a significant part of her job involves trying to protect the planet from asteroids and comets. The social situation largely doesn't matter—it's difficult to bring up casually without going into detail. Hello, I'm David, I work in advertising. Hey, I'm Shreya, nice to meet you; oh, I'm a quantity surveyor. And what about you, Megan? I work out how to use nuclear weapons to prevent stuff from space crashing into us and killing a million people. Oh. Whoa, okay. That's different. Huh. Okay. And what do you do, Anya? Oh, a landlord, cool, cool stuff, probably.

"You worry a little about planetary defense, because it sounds so science fiction that you think people won't take it seriously," Bruck Syal told me, not long after DART began its journey to Dimorphos. I suspect that many planetary defenders worry that their occupation is so inherently interesting that it makes every other job seem boring. But fair enough: to many, planetary defense may sound like an imaginary role. After the initial surprise, you can explain what the risks are and what needs to be done to get those risks as close to zero as possible. "But when you say out of the blue: I work on defending the Earth from asteroids and comets, people don't know what to do with that."

Bruck Syal coauthored a 2021 paper[1] that explored the impact of the radiation unleashed by nuclear explosions. "In the future, a hazardous asteroid will find itself on a collision course with Earth. For asteroids of moderate size or larger"—our city killers and country crushers—"a nuclear device is one of humanity's only technologies capable of mitigating this threat via deflection on a timescale of less than a decade." She and her coauthors go on: "The outputs and effects of nuclear explosions are well characterized; nuclear devices are a mature technology. All of these factors make the nuclear option a prime choice for combating asteroids on a collision path with Earth." In other words, if time is short, and we need something to deliver a lot of destructive energy to an asteroid to blow it up or knock it off course, a nuke is the most effective tool we currently possess. "Anytime you look at the problem objectively," Bruck Syal told me, "it just falls out that, okay, we would really need to use a nuke for a lot of scenarios."

This conclusion isn't coming from an armchair planetary defender. Bruck Syal works at the Lawrence Livermore National Laboratory in California, just a little east of San Francisco. If you were able to meander through its grounds, you might think you were on a university campus. But one cannot simply walk into Livermore's multibillion-dollar science complex, because this is not a home exclusively for planetary defense specialists. It's America's clandestine locale for all things nuclear weapons.

Named after the physicist who invented a type of particle accelerator, the LLNL has played a key role in the county's nuclear weapons development. In 1957, for example, it conducted the first fully contained underground nuclear explosion in a Nevada tunnel.[2] The LLNL is, in its own words,[3] designed to ensure "the safety, security and reliability of the nation's nuclear deterrent." Its mission is to make sure that the nation's nuclear stockpile is kept up to date with the rest of the world's, while using science and engineering to "enhance the nation's defense," in whatever way works.

That also includes exploring nuclear fusion, a hypothetical but realizable source of abundant clean energy in which the energy-making, particle-melding process at the center of the Sun is replicated safely on Earth. The first country to come up with a viable nuclear fusion reactor will not only be able to drastically reduce their own greenhouse gas output; they will also become an energy-providing superpower. Naturally, America wants to get there before anyone else, and LLNL's National Ignition Facility is contributing to that effort by occasionally firing 192 high-energy lasers at hydrogen atoms[4] and seeing what transpires. Plenty of other scientific spells are also cast here: LLNL scientists have frequently co-discovered new elements; the facility makes aerogels, which are incredibly light but strong solids often dubbed "frozen smoke"; and its biomedical researchers even contributed to the worldwide effort to map the human genome in its entirety.[5]

So yes, the campus features more than just nuclear weapons research. But nuclear weapons and related fields remain a prominent facet of the facility. And that means Bruck Syal can only describe her work in spartan detail. "I work on a variety of national security problems, planetary defense being one of the few I can talk about in the open," she told me, after thinking on my question for a good ten seconds. I cheekily pushed her a little. "I work on the design of shock physics experiments, and some of them are driven by high explosives. We use a lot of the same methods for analysis and running multi-physics codes we use for planetary defense . . . and we can also apply those for, um, interpretation of other kinds of experiments that are addressing very urgent national security questions."

Fortunately, I wasn't there to get her into trouble. I wouldn't want to: she is a delight to talk to, another eloquent and surprisingly laid-back member of America's planetary defense team. It just so happens that someone who knows a lot about nuclear weapons knows plenty about saving the world. "I wanted to be an astronaut," she said. "But I'm really happy where I've landed. I'm okay with not being an astronaut. Well, I'd still like to be one, but it's okay for now! I'm having a lot of fun doing this sort of work."

Back in 2005, while an undergraduate studying at Williams College in Massachusetts, a scientist popped in to give a lecture about a space mission named Deep Impact. Scientists wanted to see inside a comet, so they blew a hole in the side of one to have a peek. The drama and scientific derring-do of this mission left an impression on Bruck Syal, who thought that making craters—either naturally, through asteroid and comet impacts, or via spacecraft essentially attacking comets and asteroids—was a worthwhile endeavor. And who could blame her? These impactors are the Jackson Pollocks of interplanetary artists, creating art through destructive and resplendent means. And the craters they excavate are "fundamental to understanding the history of the solar system," Bruck Syal told me.

After college, she spent the next few years at Brown University, nabbing multiple degrees while studying nearby worlds and their surplus of scars. In 2014, LLNL asked her if she was curious about using nuclear devices to stop Earth gaining a brand new asteroidal injury—and she's been working for the secretive campus ever since.

Over the last few years, during the run up to DART's launch and in the months shortly afterward, it hadn't escaped my attention that NASA spokespeople were very reluctant to speak about nuclear weapons as a possible means of planetary defense. During press conferences, the option never came up, and when reporters asked about it, nukes were largely glossed over. And, to a degree, that's fair: nuclear bombs and missiles have never really been something that people love hearing about, because their primary purpose is to kill people.

I was born in 1988, so for most of my life the Cold War—in which the planet teetered on the precipice of nuclear slaughter on multiple occasions—was largely a matter for the history books. But when the Kremlin initiated the barbaric, gut-wrenching invasion of Ukraine in February 2022, and Moscow's nuclear threats quickly became both overt and frequent, for the first time I fretted about finding myself at the bottom of a radioactive crater. The DART mission, launching into space in November 2021, overlapped with this war. It is perfectly understandable, then, that as Ukrainians fought for their right to exist against an authoritarian, nuclear-armed state, NASA did not want to talk about nuclear weapons in any manner.

Bruck Syal is fully aware of the awkward nature of nukes. "For a variety of historical and political reasons, there's varying levels of discomfort even discussing nuclear deflection or disruption," she told me. And NASA isn't alone in its squeamishness. In Japan, this is a topic of the utmost sensitivity. "Anyone who works in this area has to be aware of that—everyone's complex emotions and feelings around this technique."

The first artificially created nuclear explosion, conducted under the direction of J. Robert Oppenheimer, was the Trinity test in the New Mexico desert on July 16, 1945. Though it involved what would today be perceived as a low-yield nuclear device, the explosion swallowed the sandy earth of the Jornada del Muerto, turning it to glass and broiling the sky, creating an expanding inferno visible from 200 miles away.[6] It was as if an alien star, tinted violet and blue for a moment,[7] had been dropped from the heavens, before being consumed by that iconic mushroom-shaped cloud. To everyone's relief, the vanishingly remote possibility that the explosion would set the atmosphere on fire and destroy the entire world[8] didn't come to pass. Better yet, the Allies had beaten the Nazis in the race to make a World War–ending weapon of mass destruction. But the global nuclear arms race had only begun. This test had forcibly warped the arc of history, and nothing would be the same again. Humanity now had the means to sterilize the surface of the world completely and thoroughly.

You've probably heard of the concept of MAD—mutually assured destruction—that would result from an exchange of nuclear attacks between two atomically armed nations. The acronym is apt; it is absurd that it takes several states owning myriad nuclear weapons to prevent any of them from being used. Some of those present during the Trinity blast may have ruminated on the perverse protective benefits of MAD, but I doubt anyone thought that nuclear warheads could defend people in a more direct way—by using one somewhere deep in space, far from home. But that is precisely what researchers like Bruck Syal are working on. "There are some people that will perceive [our work] in a negative way because of the complicated history of nuclear weapons in our world," she told me. "Our approach is, well, if someone calls us and there's an emergency, we're going to feel better having done the homework at the time."

The United Nation's 1967 Outer Space Treaty states that any nation exploring space should be doing so for purely pacifist reasons.[9] Several spacefaring nations have found moon-size loopholes in this outdated text. But one of its key tenets—"States shall not place nuclear weapons or other weapons of mass destruction in orbit or on celestial bodies or station them in outer space in any other manner"—is something that, for now, nobody has any quarrel with. Indeed, Bruck Syal does not want nuclear weapons to be deployed in space. Like most planetary defense scientists, her preference would be to use a nonnuclear technology to prevent a fatal asteroid strike—"if that would be effective," she hastened to add. But this rule may mean little if an asteroid is plunging toward us. "I think in a real emergency, people's views would evolve, and quickly."

A nuclear explosion, whether it's fission-made (splitting atoms apart) or fusion-based (gluing atoms together), is not like a chemical explosion. In both cases, a lot of energy is released as heat and sound. But nuclear explosions, with fireballs that can reach temperatures in the tens of millions of degrees, emit unique types of radiation, from invisible X-rays to neutron particles.[10] And it is that radiation that would help us deal with problematic asteroids.

A nuclear detonation on Earth is not the same as one in space. Obviously, it's far better to blow up nukes in space for all sorts of reasons. Chief among them: the radiation doesn't contaminate anything we need, and nobody dies. But in the depths of space, there is no atmosphere for a nuclear explosion to push out of the way. There is no mushroom cloud. Instead, there is a paroxysm of nuclear radiation.

"It would look like a bright flash," Bruck Syal told me. Ideally, the detonation would happen "many millions of miles away from us, so it wouldn't be visible from the Earth." When set loose upon an asteroid, much of that energy would get absorbed into the rock. Within a few millionths of a second, the radiation-suffused rocks would shatter—and that is the power that can be used to kill an asteroid in one of two ways. The first way, and the one most familiar to many, is disruption: the act of pulverizing an asteroid into a scattered and expanding cloud of tiny pieces of debris, 99.5 percent of which would never reach Earth.

And this is where *Deep Impact*, the second space-object movie to be released in 1998, comes in. This one, essentially a remixing of the novel *The Hammer of God* by Arthur C. Clarke, features a plot wherein nuclear weapons are used to blow up an impactor—this time, a comet—from the inside out. The asteroid in *Armageddon*, apparently the size of Texas, is hilariously massive, pretty much the size of a small dwarf planet. The comet in *Deep Impact* is comparatively puny, about seven miles long—enough to kill billions of people, but much easier to stop with a handful of nuclear warheads. In *Armageddon*, Bruce Willis and company only has eighteen days to save the world. In *Deep Impact*, the world has a whole year to act before impact. A veritable cakewalk. What could possibly go wrong?

Quelle surprise, everything goes wrong. The plan is to land a crew—a mixed American-Russian team of astronauts—on the incoming object, drill a bunch of holes into it, drop in a few nuclear warheads, fly to a safe distance, and detonate. Unlike *Armageddon*, the plan in the *Deep Impact* movie isn't to cleave the space rock in two, but to vaporize the cometary matter and use the explosion to push it out of the way—to deflect it. Instead, they poorly disrupt it, carving off a small shard that hits Earth and kills millions of people, with the heftier planet-killer chunk still heading toward the planet. That bigger piece is ultimately destroyed when the crewed spacecraft flies into a cavern on the planet-killer's side and triggers the remaining nuclear warheads, turning the comet into gelid confetti. Still, the original plan accidentally caused the worst disaster in human history.

At the movie's end, the US president—played by Morgan Freeman—gives a hope-filled speech in front of the trashed Capitol Building in Washington, DC, declaring that "cities fall, but they are rebuilt; heroes die, but they are remembered." The entire film is less of an action thriller and more of a planetwide drama. As that final monologue suggests, its creators were going for gravitas. Nobody comically rides a nuclear bomb.

But in real life, no heroes have to die. If you're going to use nuclear bombs to stop an asteroid or a comet, the very last thing you would try to do is land anyone on it and start carving out subterranean homes fit for weapons of mass destruction. "It is true that a nuclear explosive would couple more of its energy to an asteroid if you buried it," Bruck Syal told me. "But the operational complexity of rendezvousing, and landing, and then drilling is . . . all of that would slow you down so much that it's hard to conceive of a scenario where you'd prefer to do that just versus a stand-off explosion."

A stand-off explosion is when you detonate the nuclear device close to the asteroid's surface, at the point at which the radiative attack would cause the most damage and thoroughly fragment the entire body. There would be no need to risk any spacefarer's life. You would mount a nuclear device to an uncrewed spacecraft, send it to rendezvous with the asteroid as far in advance of its scheduled meeting with Earth as you can, and press that red button from the comfort of your mission control room.

To test out the effectiveness of nuclear disruption, Bruck Syal and her colleagues have run countless simulations on a range of impact scenarios. In 2021, she was the coauthor of a study[11] that flung various rocky hostiles at the planet along a range of trajectories, sometimes with plenty of warning time, often with barely any time to react at all. They found that if a 330-foot asteroid—the sort perfectly capable of destroying a city or a decent patch of an entire country—were heading our way, we could use a one-megaton nuclear warhead to successfully disrupt it. If it was hit at least two months prior to impact, 99.9 percent of its mass could be blasted out of Earth's way.

This, however, would be the civilizational equivalent of a Hail Mary, a last-ditch attempt to stop the apocalypse. Given the choice, nobody would want to wait just months prior to the asteroid's impact to try and prevent it. And there is no guarantee that disruption would work. Instead of fully fragmenting the target, turning a cannonball into tiny ball bearing–like pieces that would harmlessly ignite in our atmosphere, the nuclear explosions might transform the projectile into a shotgun spray of mass murder, a family of city killers that still hit Earth and rack up millions of casualties. Cities can certainly be rebuilt, but it makes it a lot harder if the craters left behind by these impactors are also plagued by radiation.

And despite our species' predilection for crafting increasingly lethal manners in which to destroy each other, there is currently no launch vehicle that could carry a powerful nuclear weapon into deep space. It isn't unrealistic to imagine a future in which such interplanetary missiles exist, although it's far likelier that the nuclear device would first be shepherded into space via a rocket, which would then find its way to the asteroid in the mechanical stomach of an uncrewed spacecraft. But nobody wants to live in a society in which nations have that capability—not least because it may start a new type of nuclear arms race, but also because accidents happen. Rockets explode on the launchpad, in the sky, or on the boundary of that thin blue line that protects us all from the vacuum of space.

Nuclear warheads have already been sent to the edge of space. The Soviet Union conducted extremely high-altitude nuclear weapons tests, but America quickly took the ignominious top spot in this competition. On July 9, 1962, an American weapons test known as Starfish Prime detonated a 1.4-megaton thermonuclear bomb more than 250 miles above the Pacific Ocean—about the same altitude that the International Space Station hangs above the world. It caused streetlights to malfunction in Hawaiʻi, created a shimmering auroral glow that could be seen as far away as New Zealand, and turned a pitch-black night for some skywatchers into an artificial noon.[12]

If the world had no choice, and it had to either risk launching nuclear weapons into space to stop an asteroid impact or just take the hit, odds are that those nukes are going to fly. But if something were to go wrong, not only would the effort fail to stop the impact, but it would spray radioactive matter across the launch site or, should the rocket fail at a significantly higher altitude, over a much larger region. The only thing worse than a city killer or country crusher asteroid hitting the planet is a city killer or country crusher hitting a planet that is already experiencing a sequel to the Chernobyl disaster.

Nuclear disruption is also something that only works—or, more accurately, might stand a chance of working—for those modestly sized asteroids. In 2019, another team of scientists simulated throwing smaller asteroids at significantly larger ones, the sort capable of killing every living thing on the planet. Although the aim of this study was to test the disruptive effects of a massive nonnuclear impact on an incoming asteroid, the extreme energies involved in these virtual runs were not dissimilar to those that would feature in a stand-off nuclear explosion.

What they learned was disquieting. Yes, the target asteroid—in this case, one just over fifteen miles wide—would shatter, sending Ping-Pong-sized balls flying in all directions. But this debris, seemingly destined for interstellar space, soon began to decelerate, before coming to a stop. Then those pieces crawled back toward the still-intact core of the asteroid, agglomerating onto its surface like frog spawn. Like the liquid metal assassin in *Terminator 2,* this asteroid was sufficiently massive that, using its own gravitational field, it could speedily reassemble itself.[13]

Ah. So that's what the smartest man alive meant in *Armageddon* when he said: "You could fire every nuke you've got at her and she'd just smile at you and keep on coming." I'm sorry I ever doubted you.

Olivier Hainaut, our asteroid explainer from the European Southern Observatory, sums it up best: "Rule number one: never break the asteroid." Too many things could go wrong with a technique so closely associated with desperation. "What you want to do is deflect the asteroid." It is possible to kill an asteroid with nuclear warheads. But directing it onto a path that doesn't end in its own destruction is a materially more sensible option. You don't want to kill an asteroid if you can help it. Ideally, you want to kill its chances of hitting us, whether you use a kinetic impactor, like DART, or deploy a nuclear warhead.

Hypothetically, a kinetic impactor spacecraft can impart a hardy bop to a midsize asteroid. But if we do not have much warning time, or if the asteroid is more of a country crusher than a city killer, we would want to throw as much energy at that asteroid as possible, to give it less of a nudge and more of a shove in the right direction. Place a nuclear warhead on a spacecraft, park it as close to the asteroid as possible to ensure maximum radiation absorption, and fire away.

In this scenario, nukes are our friend. When browsing your local supermarket for asteroid-punching bombs, your choice of nuclear device really matters. Sure, you want the explosion to have considerable oomph, but you must consider the flavor of the device's radiation. For many nuclear devices, Bruck Syal explained, "X-rays provide most of the impulse." X-ray radiation would shower the asteroid's surface but also quickly make its way below the target's lithic skin, turning solid matter into vapor. Great as that is, neutron radiation, produced by other types of nuclear weapons, may be better.[14] X-rays would penetrate no more than an inch into the asteroid, but neutrons could get down to tens of inches. "They're heating up a lot more mass that then is vaporized and expands off the surface," said Bruck Syal.

Either way, converting enough of the asteroid's face into vapor in effect creates a rocket booster. "The rest of the asteroid moves in the other direction while this material is expanding at many kilometers per second," she said. Contrary to what movies often depict, from flashes that build into brilliance to eruptions of luminous matter, this explosive process happens faster than the time it will take you to read the rest of this sentence. "Typically, that blowoff, the momentum, that's all imparted over the first few hundred microseconds."

There is likely a fine line between nuclear deflection and an inadvertent nuclear disruption—something that *Deep Impact* illustrates (despite the harebrained way in which they go about it). No matter which type of nuclear right hook you choose to throw, "you want to make sure it's not fragmenting the asteroid," said Bruck Syal. The same risk applies to a DART-like kinetic impactor mission. Hitting a city killer–sized asteroid too hard could shear off dangerously massive fragments that continue to sail toward Earth. Saving the world, whether you're relying on X-rays or metal boxes with a death wish, will never hinge on making the biggest boom. The sledgehammer will rarely turn out to be the best tool in the box. Hope should be placed in the surgical use of a chisel—or several chisels, depending on the nature of the threat.

In an optimistic timeline, a DART-like mission could be used to deflect a genuinely hazardous asteroid off course. But even then, the fight will be far from concluded. City killer asteroids are problematic enough, but what about those country crushers? Take an asteroid that's 1,600 feet long; it wouldn't be the end of the world as we know it, but it could cause continental-scale devastation. Could a kinetic impactor prevent an asteroid of that size from impacting Earth? Bruck Syal has run the numbers, and fortunately, the answer is yes—but it's a very conditional yes.

Oddly, this is related to NASA's designs on the Moon. The agency is currently working on the sequel to its Apollo program: Artemis. Should all go swimmingly, it will see America return astronauts, including the first female astronaut, to the lunar surface by the decade's end, marking the start of a permanent US presence there. NASA has been increasingly leaning on the comparatively cheap, reusable, and reliable launch vehicles developed by SpaceX for a variety of space exploration ventures. But its Artemis program relies on the success of its own homegrown Space Launch System, a colossal, record-breaking rocket capable of producing 8.8 million pounds of thrust.[15]

If a 1,600-foot asteroid was going to impact Earth, and the world balked at the thought of sending nuclear-armed rockets through the atmosphere, then America could opt to use something like the Space Launch System. According to Bruck Syal, for an asteroid that size, a successful deflection would require the use of four of these rockets, all launching without a hitch, all making their way through deep space, and all hitting their bullseyes on the asteroid in perfect succession. And we would need at least twenty-five years of warning time—anything less, and the asteroid would not be deflected quite enough, resulting in an impacted Earth.

Bruck Syal, who is also a member of the DART investigation team, reiterates that a kinetic impactor can be "a great first choice. But as the time gets short, and as the asteroids get bigger, your options are much more limited." The math doesn't lie—certain planetary defense situations can "only be addressed with a nuclear explosion."

Using a nuclear warhead does not mean your mission design is inflexible. When given short notice, the best option could be to fly out to greet the asteroidal antagonist and immediately blow up the bomb. But with greater lead times, a nuclear device could be mounted inside an AIM-like scouting mission, one that could reach the target well in advance and examine it up close, ascertaining its Earthbound trajectory with great precision while also working out if a kinetic impactor would succeed in deflecting it. If it was determined that a DART-like mission would not impart enough momentum to perform a successful deflection, then the nuclear device on board the scouting spacecraft could be triggered instead.

Perhaps kinetic impactors will be all we need to defend the planet for centuries to come. But we could also live to see the strangest of days, when a nuclear weapon is used not to threaten or end millions of lives, but to save them. I wonder how Oppenheimer would respond to such irony. Reacting to the Trinity test, he famously quoted from the Bhagavad Gita, a sacred Hindu text: "Now I am become Death, the destroyer of worlds." As noted by *Wired*,[16] this choice makes it seem like Oppenheimer (perhaps regretfully) considered himself to be an avatar of death—but only when taken out of context.

The Bhagavad Gita, starring a warrior prince named Arjuna and the deific Lord Krishna, does traffic in visions of annihilation. But it also threads in notions of purpose, duty, and heavenly supremacy. Oppenheimer's quoted line was spoken by Krishna, who becomes a staggering, boundless, multi-eyed, many-mouthed entity. Arjuna, the soldier, does what he can to affect the outcome of battles, but as is made clear by that awe-inspiring transmogrification, the fates of the many are ultimately shaped by the empyreal. The father of the atomic bomb extracted several meanings from this text, and the sight of the Trinity test understandably made Oppenheimer think of that indelible image of Krishna. But perhaps he also compared himself in some ways to Arjuna: someone performing their warrior-like duty to help the good and noble people of the world triumph over evil, whatever it takes.[17]

The thought of using a nuclear weapon in any situation, even to defend the planet against the forces of cosmic indifference, is an unsettling one. The consequence of a successful world-saving campaign, though, would be no different to using a kinetic impactor: in both cases, our future would no longer be determined by what happens in the heavenly realm. Humanity would have seized control, made its own choice to fight against the firmament, and done its best to survive. Oppenheimer and his retinue, in their quest to usher in an end to the bloodiest conflict of all time, may have offered Earth a method of complete self-destruction. But they also gave us an instrument that may forestall a celestial extermination.

Unlike kinetic impactors, Bruck Syal is confident that we don't have to rehearse the nuclear deflection or disruption technique. The Outer Space Treaty would make such a trial run illegal anyway, but the science behind this type of planetary defense method is steadfast—unfortunately, in some respects, because so many nuclear weapons tests have taken place in the last seventy-eight years, particularly during the Cold War.

"We don't think we need to do a test," she told me. "We have a very rich nuclear test history in the US, and we leverage that really heavily" when simulating asteroid deflection or disruption scenarios. Nuclear radiation, including showers of X-rays and neutrons, can also be produced at research facilities all over the world, including at LLNL's National Ignition Facility. Place an asteroid-like rock in there, fire some radiation at it, see how much of its surface gets ablated, and you've got yourself a simulacrum of a stand-off nuclear blast in space. "No one I know is advocating for a full test," said Bruck Syal. "Unless it was really necessary, you'd prefer to avoid that."

Even if the support for such a test does eventually grow, there is no dimension in which astronauts land on an asteroid, drill holes into it, and drop nukes down them. Like many aspects of these flamboyant flicks, *Armageddon* and *Deep Impact*'s world-saving nuclear schemes are entirely divorced from reality. Yet most astronomers view these movies less harshly than you may expect.

The Chicxulub revelations of the 1980s, and the Shoemaker-Levy 9 collision with Jupiter in 1994, raised the possibility of present-day lethal impacts to many in the scientific community. But in 1998, two seriously flawed but visually spectacular blockbusters introduced millions of rapt moviegoers to the concept of planetary defense. And for that, the head of NASA's Planetary Defense Coordination Office is grateful. "It definitely brought awareness that there was such a hazard in space that needed to be thought about," Lindley Johnson told me. "I credit both *Deep Impact* and *Armageddon* as bringing this into the public realm of attention, and it wasn't just astronomers thinking about it."

Bruck Syal doesn't consider the movies' impact to be especially negative, either. Just one thing, though: don't call the devices nuclear weapons. That implies their use is offensive, violent, for the purposes of warfare. The community instead refers them as nuclear explosive devices, or NEDs. NED may sound more like a phlegmatic work associate than a planet-protecting hero, but hey, a NED may save us all one day. And NEDs comes in different varieties. If you're using a fission device, not a fusion device, then you are opting for a MUFN—Mitigation Using Fission Nuclear device—which is, gloriously, pronounced *muffin*.[18] I can easily picture the scene unfolding in the Oval Office. The president has just been briefed on the asteroid threat by his ashen-faced security council. Upon asking what can be done to defend the planet, someone will have the unenviable task of strenuously advising that there is only one sensible course of action: we must launch a muffin named Ned into space.

NED. Really? "It's fine, but . . . yeah," said Bruck Syal, smirking.

The ominous nature of the technology Bruck Syal works with appears to do nothing to dilute her breezy disposition. Does the solemnity of her work, and the efforts of her world-protecting friends, ever gnaw at her? The year-to-year risk of a city killer hitting Earth is low, so she is never acutely anxious. "But there's just so much unknown right now that we shouldn't be totally comfortable with the status quo," she said. "I definitely have a sense of urgency in our work. If we fail to consider the right things, we could put people in danger." Ruminating on nuclear weapons—sorry, NEDs—is just another way to be proactive on a planetary scale.

In any event, Bruck Syal was more than happy to be part of the populous DART family—thrilled that the spotlight was, for the time being, on a nonnuclear spacecraft heading toward its doom. "It's incredibly urgent work, because we don't know when we need to be ready," she said. "So much is resting on the shoulders of the people who work on this."

December 2021. DART has left Earth behind. On December 7, about 2 million miles from home, it opened its DRACO eye and took its first look at the universe: a dozen stars, the intersection of several constellations, glinting against a jet-black canvas.[19]

Elena Adams and her team of engineers checked the spacecraft's subsystems from the Applied Physics Laboratory's mission control center. Everything looked normal. The celestial sailors at NASA's Jet Propulsion Laboratory confirmed that DART was heading in the right direction. Right on schedule, the team flipped the switch on the experimental NEXT-C ion thruster, giving the spacecraft a little kick in the pants. And to everyone's beady eyes, the test appeared to go well. No warning lights flickered on. For the next few months, the world's first planetary defense test looked set for smooth sailing.

But all was not well in the deep and lonely dark. DART felt a surge of electricity course through its viscera, its very own *Star Wars*-esque, R2-D2-getting-zapped moment. Something had just gone wrong—and, back home, nobody noticed.

The broken Shoemaker-Levy 9 comet heading for Jupiter. *(NASA, ESA, H. Weaver and E. Smith (STScI) and J. Trauger and R. Evans [NASA's Jet Propulsion Laboratory])*

The Chelyabinsk Meteor. *(Alex Alishevskikh via photo.alishevskikh.com/about)*

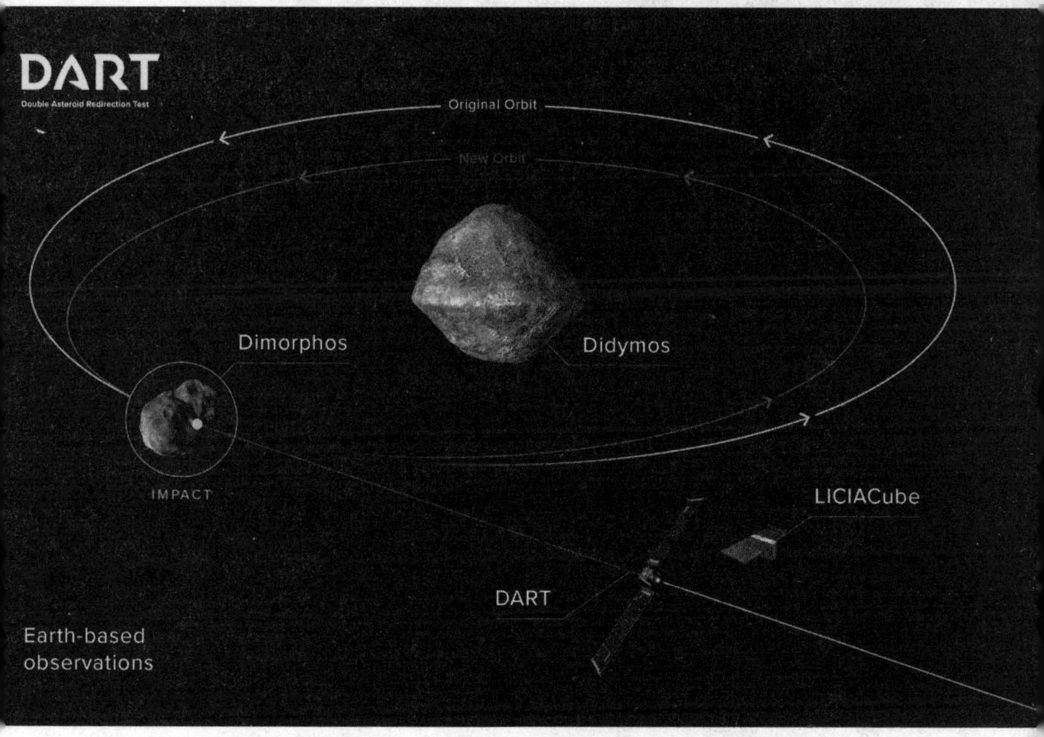

DART
Double Asteroid Redirection Test
Original Orbit
New Orbit
Dimorphos
Didymos
IMPACT
LICIACube
DART
Earth-based observations

An illustration of how DART will impact Dimorphos, deflect it, and change its orbit around Didymos. *(NASA / Johns Hopkins APL)*

Comet Tempel 1 is shot at by Deep Impact. *(NASA / JPL-Caltech / UMD)*

Comet 67P/Churyumov-Gerasimenko outgassing. *(ESA / Rosetta / NAVCAM)*

The toaster-size LICIACube satellite arrives at the Applied Physics Laboratory, ready to be installed inside DART. *(NASA / Johns Hopkins APL / Ed Whitman)*

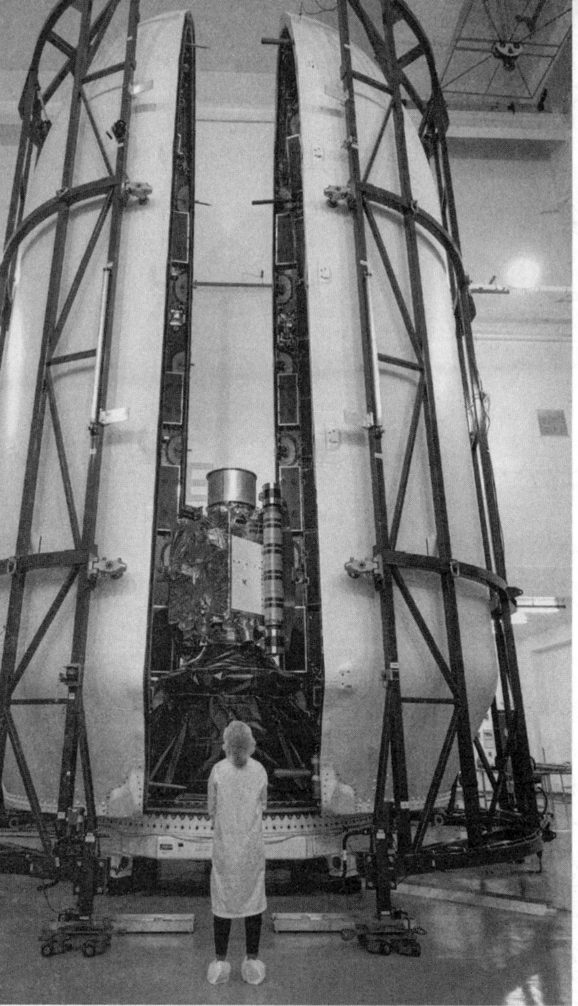

Inside SpaceX's Payload Processing Facility at Vandenberg Space Force Base in California, both halves of the Falcon 9 rocket's protective payload fairing move toward NASA's Double Asteroid Redirection Test (DART) spacecraft on Nov. 16, 2021.
(NASA / Johns Hopkins APL / Ed Whitman)

DART launches from Vandenberg Space Force Base in California atop a SpaceX Falcon 9 rocket.
(NASA / Bill Ingalls)

IV

Spying on Heaven

For David Rankin, Saturday, November 19, 2022, started off like any other night. It was already cold in Arizona by that time of year, but atop Mount Lemmon, the highest point in the Santa Catalina Mountains, it was considerably chillier. Though observatories aren't always the most efficiently insulated places, Rankin was comfortable enough. By this point, he had been with the Catalina Sky Survey for four years, and as a member of the night crew, he had found his rhythm directing the Mount Lemmon observatory's 1.5-meter telescope, hoping to spy brand-new asteroids glinting in the shadows.

Discovering asteroids was quotidian. To have a week in which no new entries appeared in the already densely populated catalog of NEOs would come as a great surprise. Even the sort of rocks astronomers would have once considered miniscule, and thereby almost invisible to telescopic eyes, were now being regularly spotted by Earth's asteroid-hunting spotlights. And pretty much all of them were found to be on harmless journeys, passing by our planet—sometimes between Earth and the Moon—and making their way, like all of us, around the nuclear furnace at the heart of the solar system.

But that night, almost right at the start of his shift, a small spark showed up in the west of the sky during the telescope's preplanned survey. It was bright, but distant, a bit like seeing the headlight of a car on a darkened road across a forested valley, or a lighthouse on an island across a vast, blue-black bay. "It was moving fairly quick, and it got our attention immediately," Rankin told me. "It was making a hard turn to the left."

Whatever it was, it was traveling extremely close to Earth. It had to be one of two things. It could very well be the reflection off an artificial satellite; today, internet-providing satellites have become as common as the visible stars, and they often hug the horizon, bouncing bright sunlight into the eyes of irritated astronomers. But this object appeared to be making a beeline for the planet itself. "The second it came up on the screen, I knew it was something interesting," said Rankin. He junked the observatory's preloaded survey plan and pointed the telescope at the mysterious object, tracking it as it slinked across the sky. There was little doubt about it: he had just found an asteroid. And maybe, just maybe, it was heading toward Earth.

This is the scene in movies where the astronomer's jaw drops, they utter an expletive or say, "Oh my God," then flump back in their chair, weigh up their options, and furiously try to work out who they are supposed to inform first. Panic sets in. In a matter of months, weeks, or even days, Earth will be hit by an asteroid, and everybody is going to die. Rankin didn't know it yet, but his newly discovered asteroid was indeed plummeting toward Earth, and in just three-and-a-half hours, it would impact somewhere close to the American-Canadian border. And he was beside himself with excitement.

That's right: that moment in movies when skywatchers first spot Earth-bound asteroids does happen in real life. Finding a new NEO, though, tends not to cause astronomers to fall from their chairs; they are so brutally efficient at finding them these days that almost every discovery is anything but shocking. And although today's asteroid-hunting observatories are far more mechanically complex and sophisticated than their antediluvian telescopic predecessors from centuries past, the fundamental science remains the same.

Eric Christensen, the principal investigator at the Catalina Sky Survey, explains that like many digital cameras, optical telescopes—those that soak up the type of electromagnetic radiation visible to humans—have mirrors.[1] Those mirrors collect and focus starlight. "And the bigger your mirror, the deeper you can see—the farther you can see, or the fainter you can see," said Christensen. In other words, with a large mirror, you can detect fainter objects (those that are smaller and reflect less light overall) or bigger but darker and poorly reflective (those that are farther away from Earth). Large mirrors are fabulous for asteroid spotters.

"The other thing is field of view," he added. Field of view is the total area your camera can see. The same applies to your eye. Look up from this book right now, and stare forward; try not to freak out the person sitting opposite you, if possible. Everything you can see without moving your eyes to focus elsewhere? That's your field of view. Telescopes have them, too; sometimes wider, sometimes narrower.

Ideally, an asteroid hunter wants a telescope with a huge mirror and a vast field of view, so they can spot fainter asteroids far from home across a huge patch of the night sky. But most observatories cannot have their cake and eat it too: engineering restrictions mean that individual telescopes tend to have only a big mirror or a wide field of view. Can you ever have both? "It's not impossible, but it gets very expensive, very quickly," said Christensen.

Optical telescopes are how NEOs are usually first found. A reflective thing moving across the sky? It's probably an asteroid. Hoping to find every single NEO in case one of them decides to pay us an unwelcome visit, telescopes across the world spend some of their observing hours—or, if their mission is asteroid exclusive, all their observing hours—searching for them.

NASA supplies funds to several asteroid-hunting surveys through its NEO Observations Program,[2] and they are all remarkably good at their job. The University of Hawai'i's Panoramic Survey Telescope and Rapid Response System, or Pan-STARRS,[3] is a pair of huge and highly advanced digital cameras sitting atop the enormous Haleakalā shield volcano on the island of Maui. That same institution also runs the Asteroid Terrestrial-impact Last Alert System, or ATLAS,[4] which operates two telescopes in the same archipelago, another in Chile, and a fourth in South Africa. Software programs, tweaked by astronomers, automatically scan the stars several times a night, hoping to spot NEOs.

Together, these and other NASA-funded surveys have detected oodles of new near-Earth asteroids since the turn of the millennium. Since 2013, they have found at least a thousand new space rocks every single year, and "the current crop of surveys—almost all NASA-funded, ground-based surveys—are finding about 3,000 NEOs per year," said Christensen. For the past two decades, the Catalina Sky Survey often achieved the highest annual tally.[5] It is an asteroid discovery machine, one that's come a long way from its humble origins as an undergraduate project in the 1990s. Back then, students wondered if they could stand atop the Santa Catalina Mountains to find asteroids. So they gave it a go, manually guiding their telescopes and using old-school wet chemistry to develop photographic plates, hoping to spy those telltale blips of light. And it worked. "There were big, fat, juicy near-Earth asteroids just waiting to be discovered," said Christensen.

After proving the concept, NASA funding began to flow, and various telescopes began to be modernized. Today, the Catalina Sky Survey's three telescopes in the Santa Catalina Mountains—two at the summit of Mount Lemmon and one atop Mount Bigelow, all owned and managed by the University of Arizona—have but one purpose: to find NEOs before they find us. "We don't have to share the telescope time with any other users," said Christensen.

If you were to accompany an astronomer during one of their world-protecting observation shifts, drama would be unlikely to unfold. The telescope moves largely by itself, following a program designed by its masters. If something curious is identified, it alerts the nearest human, who carefully checks the source of the signal and determines whether it's noise or something real, and what the source may be. "They are usually up all night, and they are usually working alone," Christensen explained. Some play video games. Some read books or work on papers.

I've even seen night-shift astronomers take little naps, because why not? Once, a scientist I was with was explaining the wizardry behind his telescope when he accidentally knocked his smoothie onto a dashboard. As he fretted about cleaning it all up, I was both comforted and a little disappointed that the telescope didn't start spinning around wildly. Things tend to be quiet. Unless, of course, you discover a new asteroid, and that asteroid is going to crash into you in just a few hours' time.

Here's where movies mess up. The astronomer that first sees and tracks a possibly Earthbound asteroid does not then start doing mathematical gymnastics on a notepad or a whiteboard, praying to the ghost of Copernicus that it will reveal a close-to-zero possibility of imminent impact. Observatories can do this, but they use computer programs to work everything out. Rankin ran calculations at Mount Lemmon the night he found his sketchy-looking asteroid. "But we defer to Scout, because they're the experts," he told me. "And Scout latched on to it immediately."

Whenever an astronomer anywhere on the planet catches something novel spinning in the heavens, their observations are relayed to the Minor Planet Center in Cambridge, Massachusetts, a NASA-funded clearinghouse that keeps a record of asteroids, comets, and anything that isn't a full-blown planet or one of its natural satellites. Anyone, from professional astronomers to fascinated amateurs, can access this data to see what ballets are happening above our heads. But one of the primary recipients of such observations is Scout—a collection of algorithms housed at the Jet Propulsion Laboratory's Center for Near Earth Object Studies, or CNEOS. If NASA's Planetary Defense Coordination Office had a map room, it would be CNEOS. And Paul Chodas, its director, would be the cartographer-in-chief.

When we chatted, I accidentally caught him at an inconvenient moment. "I'm a little bit distracted because the US soccer game is underway," he told me, looking offscreen at the tiny figurines wandering about on the pitch during the 2022 World Cup. "It's halftime, so I do have a little time." (The US beat Iran, 1–0, readers.)

Fortunately, I had just enough time for him to get me up to speed on the history and purpose of CNEOS. Its job is to keep track of every single suspicious-looking asteroid and comet close to Earth's doorstep. Its staff are the watchers on the wall, the searchlights atop the towers. And they haven't been around for too long. In the early 1980s, the Jet Propulsion Laboratory was tasked with charting accurate orbits for all the solar system's paraphernalia, including Halley's Comet,[6] which made a close approach to Earth in 1986. The orbital paths of many mini-worlds were traced out by software coded by Chodas, the primary objective being to make sure that the NEOs among them were not coming our way anytime soon.

When Shoemaker-Levy 9 was discovered, Chodas and his colleagues were asked to confirm whether it would impact Jupiter. They ran the numbers. "It went to 100 percent very quickly," he said. That they also calculated where on the gas giant each individual fragment would impact was an impressive flourish. "The first impact, Fragment

A, hit. It was observed, and we were so happy. It wasn't a fizzle; it was a grand event." Some of the Shoemaker-Levy 9 impacts also happened during the final match of the 1994 World Cup, with victorious Brazil planting deadly penalties in Italy's net as a planet killer comet scored a total of twenty-one epic goals in Jupiter's stormy skies that same week.

With planetary defense now emerging as a top-tier issue, NASA made Chodas and company the official group that would tell the country, and possibly the world, how safe (or not) everyone was from all those asteroids and comets whizzing about. "We had that expertise, so they formed the Near-Earth Object Program Office at JPL in 1998," he said. At the time, they had meager computing capabilities—"We could do one object at a time"—but now, thousands of orbits can be calculated simultaneously. When the Planetary Defense Coordination Office was inaugurated in 2016, Chodas' unit was christened CNEOS.

Davide Farnocchia is one of his cartographers. He is a navigation engineer, someone who can work out whether a spacecraft like DART is sailing through the stars in the correct direction. These same principles of navigation can be applied to asteroids and comets, even though humans have no influence over their movements. "They just do their own thing, and we try to understand what they're doing," he told me.

In 2015, he decided to bundle a bunch of code together and name it Scout. This little critter gobbles up any new observational data that has been newly posted by the Minor Planet Center and does its best to connect the dots. Astronomers prefer to track an object by telescopes all over the world, for as long as possible, to precisely plot out its orbital highway around the Sun. But if an asteroid is first found when it's practically ringing our planet's doorbell, it would be nice to know, based on a very small number of observations, if there is a chance it could crash into us in the next month.

The overwhelming majority of objects assessed by Scout do not raise any alarm bells. But on November 19, 2022, things played out very differently. Rankin sent his smattering of observations to the Minor

Planet Center, and Scout dutifully ate them up and digested them. Just forty-five minutes after Rankin had made his discovery, Scout sent him an email: there was a 25 percent chance the asteroid would enter Earth's atmosphere anywhere from the Atlantic Ocean off the eastern seaboard to somewhere around Mexico.

The call was put out to nocturnal skywatchers around the world, and many answered. "The probability jumped to 50 percent. That's when I really started getting excited," Rankin recalled, coming close to gleefully rubbing his hands together. His dream of discovering an asteroid that was about to slam into the world was bolstered by a group at the Farpoint Observatory in Eskridge, Kansas. After they tracked the asteroid over the next hour,[7] Scout came to a confident conclusion: at 3:27 a.m. Eastern Standard Time, the asteroid was going to careen into southern Ontario. The appropriate Canadian authorities were informed, and, as the planet braced for impact, Rankin sat back in his chair inside his mountaintop base and began to cackle maniacally.

All right, I'll come clean: If you are wondering why you don't remember Canada being wiped off the map in 2022, that's because it wasn't. And as far as I am aware, Rankin did not laugh like a supervillain. An asteroid did pay Ontario a visit that night—but based on how much light it was reflecting, everyone knew it wasn't a threatening monster, but a pip-squeak, just three feet long. Unless there happened to be an unfortunate astronaut hanging out in low-Earth orbit right in its path, this tiny asteroid was never going to harm anyone. The event, however, proved to be a fantastically successful test of NASA's telescopic eyes and cybernetic cartographers.

Exactly as Scout foresaw, the asteroid perished at 3:27 a.m. Coincidentally, an astronomer working for the Pan-STARRS survey lived right below the atmospheric entry point. Rankin quickly sent him a message and told him to watch the skies, just in case this asteroid's demise came with any fireworks. The moment he stepped out of his home in London, Ontario, "the fireball went right over his house!" An aquamarine-tinged streak of light streamed toward the horizon, and "it caused quite the sonic boom when it came in."

Many people have had a chance to witness a shooting star. I've been lucky enough to stand at the summit of Japan's Mount Fuji, a slumbering volcano built from ancient frozen fires, at the peak of the Perseids, a periodic meteor shower given to us by the comet Swift-Tuttle. A gift of cosmic serendipity, our planet passes through its tail at the same time every year, and its discarded wisps burn up in our atmosphere, flooding a night sky unpolluted by artificial light with diamantine tears. It is, without question, one of the most beautiful things you can ever lay eyes on. Such an experience is accessible to all those willing and able to make the ascent. But finding a solitary asteroid moments before it meets its end, knowing when its spectacular immolation will occur, and witnessing it ignite (in person or remotely via recorded footage) in Earth's skies, is a considerably rarer marvel: Rankin was just the sixth person in all human history to have this exclusive experience.

That doesn't mean Earth-impacting asteroids are rare, though. A miliary defense system designed to spot rogue nations secretly testing nuclear weapons reveals that we are hit by similarly sized objects all the time. In 1996, the UN adopted the Comprehensive Nuclear-Test-Ban Treaty. This treaty prohibits the creation of any nuclear explosions anywhere for any reason, regardless of their yield. Several nations possessing nuclear technology have yet to ratify the treaty, meaning it cannot currently be legally enforced. But understandably distrustful of certain authoritarian regimes, and keen to assess compliance to the treaty by both signatories and nonsignatories alike, the UN uses a monitoring network that has mechanical ears all over the globe.[8] For example, seismometers, primarily designed to detect earthquakes, can also pick up on vibrations produced by nuclear blasts, as can underwater hydrophones that monitor the symphony of the world's seas. Infrasound (low-frequency sound that's inaudible to humans) sensors can detect big booms, too—and these sensors are especially good at hearing the death screams of meteors blowing up in the sky.

The treaty's multi-eared network has spent most of its life listening to the noise of North Korea's nuclear weapons tests. But in conjunction with an assortment of weather satellites and various US military and intelligence assets, it has also recorded the spectacularly fiery ends of about 1,000 asteroids.[9] Many release as much energy as a reasonably powerful nuclear weapon, but these asteroids mostly flare out above our oceans, and as such are cremated without an audience. The point, really, is that Earth is always in the firing line, but our atmosphere protects us from most of the solar system's bombardment.

The Ontario impact event was never a cause for concern. But it did serve as a reminder that the unthinking universe will eventually send something frightening our way—and we cannot afford to take our eyes off the skies. Imagine, CNEOS's Farnocchia told me, if something like Chelyabinsk or Tunguska was about to happen again over a heavily populated area. Even if the asteroid was spotted just hours before impact, Scout could sound the Klaxon. There may not be anything anyone could do to stop the impact, but it could give those at ground zero a vital heads-up. "You can tell people: if there is a flash in the sky, don't run to your window," he said. If they have time, they can flee or find shelter. "You can protect people."

To Rankin's relief, he didn't have to concern himself with such responsibilities that fateful November night. All he was worried about was that his shift peaked way too early. "It happened quick. I found it, we determined it was an impactor, it impacted, and I still had another five or six hours left that night," he told me. It's true, not many people start their work shift by finding an asteroid that hits the planet. "It was a major rush early on in the night, then it was a little bit of a winddown after that."

Right, I thought: better let him get back to it. I wouldn't want to ruin his chances of adding to the Catalina Sky Survey's asteroid tally. I asked him if regular, Earth-avoiding NEO discoveries now felt rote. Not especially, but Rankin was excited at the prospect of spying another imminent impactor. "I'm keeping my fingers crossed for a second" asteroid primed to crash into Earth, he said—before quickly adding: "Also small! Also small."

It turns out that there is little in life more reliable than orbital dynamics. Observe an asteroid or comet long enough, and its orbit over the next century—sometimes, the next few centuries—can be calculated with an extremely small margin of error. CNEOS has Scout, its short-term "are we all screwed" town crier. But if Scout finds that no impact will occur in the next thirty days, it hands things off to Sentry, our long-term prophet of the planet's condemnation.

Farnocchia, Scout's architect, also maintains Sentry. In very reductive terms, he makes sure the bulb in the giant spotlight is kept working. Sentry's job is as straightforward as it is important to all 7 billion of us. If an object is confirmed to be a real asteroid or comet, "Sentry then analyzes the possible orbits over the next one hundred years to see if there is a possible impact," Farnocchia explained. And its analyses can be found online, free for anyone to access, on the Sentry Risk List.[10] This is a catalog of NEOs that stand even a remote chance of impacting Earth within the next century. It also features a smattering of asteroids (those with extremely precisely plotted orbits) whose possible impact dates are known a little further into the future.

That you can click on any possible impactor and find out how likely it is to kill half the people on the planet—not just in any year, but on a specific day in any year over the next century—is remarkable to me. Trying to come up with a way to avoid hanging out with your extended family during next year's Thanksgiving? Desperate to avoid the baby shower of your most narcissistic work colleague? Go to the Risk List and you may find that a giant asteroid could destroy the world that same day, relieving you of your social responsibilities. Unfortunately, though, it looks like we will all be afflicted by many such gatherings for many years to come: nothing on the Risk List, for the time being, stands anything close to a decent chance of hitting Earth.

Toward the start of 2023, a newly discovered asteroid dubbed 2023 DW, one roughly the same size as Tunguska, was found to pose an impact risk. At the time of its identification, it was estimated to have a one-in-710 chance that it would hit Earth on February 14, 2046. As much as I darkly relished the thought of the corniest of all Hallmark holidays being trashed, I'm also a human being and therefore don't want a city to be wiped off the map. That is why, a week later, it was a little unnerving to see the odds of that 2046 impact rise to one-in-360. I messaged Farnocchia to ask: Um, so, when is it going to drop? "We cannot say if and when the impact probability will drop, because that depends on what the actual trajectory of the asteroid is," he told me at the time. "If DW is on a near-miss trajectory, say one lunar distance, then the impact probability will go up for a while, until the uncertainty shrinks enough and gets smaller than one lunar distance, and therefore the Earth falls outside of it." And if it doesn't do that? "If 2023 DW is in fact on an impact trajectory, then as the uncertainty shrinks the impact probability will go up until it reaches one."

The thing is, one-in-360 odds aren't that high, despite what it sounds like: that's still a 99.72 percent chance of a miss. And as you've probably guessed by now, additional observations of the object later revealed that this Valentine's Day impact stood essentially no chance of occurring.

Although it feels like it should be worrisome, this up-and-down odds rollercoaster is a common occurrence, one that almost never disturbs astronomers. Not long after an NEO is discovered, it has been observed only a handful of times, and so its orbital path is more uncertain. Imagine you're driving on a highway with 1,000 lanes, and you're close to the centermost lane. At the end of the centermost lane is a titanium wall, so anything zooming to it is certain to crash, but all other lanes are perfectly safe. With just a few observations, Sentry cannot tell which lane you are driving in, only that you aren't in the outermost lanes—so you have a small chance of impact. More data comes in, and because Sentry now knows you are in one of the central lanes the impact odds go up. A couple more observations are considered, and now Sentry knows you are a dozen lanes away from the centermost lane—and your impact odds drop to zero.

Sentry was already good at its job, but a recent upgrade has now bolstered its excellence further. Tiny as they are, photons—particles of light—can ping off asteroids and over time slowly nudge their orbits away from the Sun. Those absorbed photons also get reemitted, often at different angles if an asteroid spins about (many do), which pushes the asteroid back as they escape. These subtle maneuvers may not mean much in the short term, but they could put a once-safe asteroid on a collision course in the long run. Thanks to that recent bout of tinkering, Sentry can now calculate the future effect of these luminous altercations. Anything that stands even a really, *really* small chance of hitting Earth in the next century is kept on the Risk List until precise observations can completely rule it out.

For those of us who don't want to rely on America doing all the world-ending soothsaying, you're good, because Europe is doing its part. The European Space Agency has its own Planetary Defence Office,[11] which also has a counterpart to CNEOS: the NEO Coordination Centre, or NEOCC,[12] circa 2013, based in Frascati, Italy. Europe's organization has a principally similar role to its American counterpart, but there are some key differences. NASA has been active since the early 2000s in paying for, or partially funding, asteroid-hunting observatories. Although ESA is starting to develop its own homegrown Flyeye telescope network,[13] and owns some NEO observatories around the world, many in its contact file are independent but collaborative institutions that can be asked to provide observations upon request.

NEOCC is checking CNEOS's homework, and vice versa. America's Sentry is reinforced by Europe's ODIM (Orbital Determination and Impact Monitoring)[14] program. Scout is tag-teamed by Meerkat,[15] standing on the hill, and looking for predatory asteroids cresting the horizon. NEOCC has its own Risk List.[16] And, said Juan Luis Cano, a member of NEOCC, both groups have the same goal. "We want to remove as many objects from the Risk List as possible," he told me.

To remove everything from the Risk List would require discovering every single NEO and observing it long enough to know, for certain, that it won't hit Earth in the next one hundred years. That's a gargantuan endeavor, and even if everything is found, that doesn't rule out impacts beyond this century: NEOs making close passes to Earth will be tugged at, just a little, by our planet's gravity; and over epochal time, this could mean that an impact by one such potentially hazardous asteroid may occur far into the future. Keeping tabs on those NEOs is also part of CNEOS's and NEOCC's job descriptions—but, for obvious reasons, it's the more immediate impact possibilities that quite literally keep them up at night.

If you are curious about the odds of our planet being hit by a sizable asteroid over the next century, but you don't want to sift through a list of objects and decipher their weird-sounding probabilities, fret not: Richard Binzel, a planetary scientist at the Massachusetts Institute of Technology in Cambridge, Massachusetts, is here to help.

Binzel knows asteroids. He has one of those biographies[17] that you expect to hear hastily summarized by the president's chief-of-staff at the beginning of a movie: he was given a telescope for his twelfth birthday and, upon seeing Saturn's rings, decided planetary astronomy was all he should do; as a teenager he wrote scientific papers about asteroids that were not just advanced for his age, but for the scientific community; he interned in 1980 for the legendary Shoemakers; he's played leading roles on various deep space missions, won a bunch of awards, and—of course—discovered several asteroids himself.

His grandfather, a chemical engineer at MIT, was an instrumental influence during his boyhood. “We would sit out on his back patio and look at the stars. He would ask me questions. He would just kind of challenge my curiosity,” Binzel recalled. “The first asteroid I discovered I named after my grandfather.” (That would be asteroid 13014 Hasslacher, named in 1987.) “I'm not completely daft,” he added: “The second asteroid I discovered I named after my wife.” (That would be asteroid 11868 Kleinrichert, discovered in 1989. Asteroids, by the way, aren't always graced with real people's names; some, like asteroid 274020 Skywalker, get brilliant monikers. Others are not so fortunate: number 88705 is called Potato.[18])

Around the time *Armageddon* was released, Binzel invented the Torino Impact Hazard Scale.[19] Or, as you might think of it, the “should I keep paying my mortgage” scale. It's designed for public communication and is included on Sentry's Risk List. Not only does it go from zero to ten, but it's color-coded, from white to green to yellow to orange to red. A NEO with a Torino Scale rating of zero (white) indicates that there is an effectively zero chance of a collision, or the object is so small that it would go the way of the Ontario meteor should it ever hit Earth. As was the case with 2023 DW, asteroids sometimes come in rated at a one (green) and stay there until more observations come in. This rating means that a routine discovery shows an extremely tiny impact possibility, but this rating is very likely to fall to zero in the very near future.

NEOs at two to four (yellow) merit special attention, with four representing an object that has a one-in-100 chance of hitting Earth and causing regional devastation. If the impact day is less than a decade away, then we should drop what we are doing and focus on this threat. The orange zone, numbers five to seven, represents attention-worthy, higher probability impact scenarios that could destroy a country or continent. After that, we're into the red zone, where an impact is guaranteed. "Eight would just be something Tunguska-like," said Binzel. "And ten is a dinosaur-level impact."

To put it another way: white and green means you should continue to pay your bills (sorry). Even yellow recommends financial responsibility, but feel free to use it as an excuse to get out of that godawful baby shower. Orange, to me, suggests that you should go on that sabbatical or epic holiday you've been putting off for a while—you know, just in case. And unless we have a way to stop it happening, red means you should live a little more hedonistically. If it's a ten, then to hell with it: tell your work colleagues you've always hated them and their *extremely* loud chewing, tell that other person you've always loved them, and climb to the top of Mount Fuji to see the unobstructed stars.

The highest ranking any NEO has on this scale is a four, which was (temporarily) given to an asteroid named Apophis. Several are given a one. Most come in, then stay, at zero. Binzel designed this scale to be simple, to help people answer the question: Should we be worrying? "It's a filter. What do we pay attention to?" he told me.

Lest we forget, though, most city killer–sized asteroids have not yet been discovered. That means we have no idea what their impact odds are, or what their rankings are on the Torino Scale. It's possible that they'd all be zeros. But imagine if one of those starts as a one on the scale, but that number only rises and never falls. Imagine that this asteroid doesn't have a zero chance of impact, but 20 percent or 46 percent—or even 100 percent. Planetary defense methods cannot save the world from targets they cannot see: if the world's NEO-hunting efforts aren't good enough, then perhaps one day we will get a sequel to Tunguska, or worse.

All things considered, we should take comfort in the fact that nothing on the Sentry Risk List is, at the time of writing, above a zero on the Torino Scale. But all legitimate, scientifically rooted fears beget panic. And if you've simply googled the word *asteroid* in the past few years, I'd understand if you thought the world was coming to an end sometime next week. And the week after that, too.

Aaron Reich will never forget the time he measured an asteroid with giraffes. He could have used any metric—hippos might have worked, wolves would have been fun, and axolotls would have been just as useful. But when an asteroid was about to make a routine flyby of the planet back in 2022, his gut told him to deploy the giraffes. And so, on the June 1, as 2022 KP3 was primed to zip past us, the *Jerusalem Post* ran the headline: "Asteroid the Size of a Giraffe to Skim Past Earth this Week."[20]

Reich is a writer who loves space but hates fearmongering. That's a good combination. Tabloids around the world, particularly the notorious British ones, love to generate headlines that make anxious people feel compelled to buy papers or click links. In the last decade, online articles about asteroid flybys have been used to grim effect, making many think that we are always on the precipice of spaceborne destruction. So, thought Reich, why not put a tongue-in-cheek spin on things? Why not make it clear that asteroid flybys are commonplace and nothing to be feared, and bring them to everyone's attention in a novel way? Thus, the giraffe metric was born—and his article got clicks aplenty.

I asked him about the confusion that emerged online that June, with people wondering how exactly he was determining the size of a giraffe. He laughed and rolled his eyes. From that first piece, he put out hit after hit, measuring harmless asteroids using a massive menagerie, from pandas to platypuses. One story from 2023, in which he described a "corgi-sized meteor as heavy as 4 baby elephants,"[21] went viral on social media.

"I think the pug one might be my favorite," he told me. This was in reference to a 2023 article in the *Jerusalem Post* titled "2 Asteroids the Size of 100 Pugs to Pass Earth Tuesday."[22] The article goes on: "As stated earlier both asteroids have a diameter that at most measures to around

100 pugs in width, assuming those pugs were stacked on top of each other, paws to shoulders, forming some sort of adorable wheezing leaning tower of pugness." After noting that asteroid impacts are "far less liked than pugs," Reich wrote that "rather than one of these asteroids, you are far more likely to be hurt by an actual grumble of 100 pugs rushing towards you at much slower speeds." News you can use, as they say.

"I try to encapsulate the zeitgeist," he explained. Although not strictly animals, he compared one asteroid to Oscar statuettes to coincide with the awards ceremony. If he ever runs out of animals, he told me, he can always turn to Pokémon, all of which have verified dimensions. As a journalist, I was curious: What did his editors, who thankfully allowed it to happen, initially think of his approach? "At first they were a bit concerned, but after a bit, they happily embraced it."

Like David Rankin, Reich shares a widespread hope in the astronomic community—for one of these hypersonic space rocks to hit Earth. They only have two conditions: it must be small enough to cause no harm, and, unlike the Ontario meteor, it needs to survive its encounter with the atmosphere. It's all well and good to try to use DART or a nuclear device to deflect an asteroid, but how it responds to such an attempt depends entirely on its constitution: its rigidity, its mechanical strength, whether it is properly glued together or loosely bound by the silk of a gravitational web.

We need to get to know asteroids better. And for that, we need to find their traitors, their turncoats, those coming over to our side and offering us their secrets. We need meteors to become meteorites. And, more important, we need someone to go and find them—even if they happen to land in a treacherous place.

We need more people like Mohutsiwa Gabadirwe.

In 2018, the Botswana Geoscience Institute in Lobatse, a town on the border with South Africa, hired Gabadirwe as their meteorite curator.

The only problem was that he didn't have any meteorites to curate. He had plenty of ambition, but it was stifled by a lack of funds. In April of that year, he told his boss that if they wanted to showcase and study meteorites, then they should buy them—but, for the time being, no dice. What was a wannabe meteorite connoisseur to do?

He needn't have worried. About 22 million years ago, an enormous asteroid, now named Vesta, was hanging about in the belt. Vesta was hit by another asteroid, sending shards into the great beyond. One of those shards began its long, lonely journey to Earth. And in the end, Gabadirwe's package was just two months late—not bad for a delivery route that began millions of miles away in the shadow of Mars.[23]

On the morning of June 2, 2018, the Catalina Sky Survey saw something burning above, an asteroid plunging toward them at 38,000 miles per hour. Soon, other observatories chimed in, Scout lapped up their information like warm milk, and Davide Farnocchia, in the middle of a laid-back visit to the Ries impact crater in Germany, got a text—eventually. "We're in the middle of the Bavarian Forest with very crappy cell phone reception. And at some point, I got a text message saying there's a significant [impact] probability from a newly discovered object," he told me.

Scout kept sending messages his way, and the impact probability quickly rose to 100 percent. "We realized right away it was small," he said—about five feet long, based on how bright it appeared—so that was nice. And Scout predicted that it would hit the atmosphere above southern Africa. So it came to pass: seventeen miles above the ground, in the dark of night, the equivalent of 200 tons of TNT exploded in a brief flurry of light and noise. The word went out, but as it was a Saturday, nobody was listening.

"The warning did not reach us. That evening, I was out of the city," Gabadirwe told me. He had been making a fire, and just as he lit his match, another flash crept into the corner of his eye. What the heck was that? Then, his phone rang. It was his brother. "Did you see the meteorite?" he exclaimed, telling the curator that it looked like a very big star. "Everyone saw it, all over," Gabadirwe recalled, chuckling; the only exception seemed to be himself.

On Monday, a group at the institute discussed what to do, as did their colleagues at the Botswana International University of Science and Technology. Sure, cool fireball and all—but did any of it survive to make it to the ground? And if so, where in all of Botswana would it be? "That's when everything started," said Gabadirwe. Texts. Emails. Calls. A flurry of noise and action.

Peter Jenniskens joined the fray and offered his services. Aside from his knowledge of airbursts, like those that had occurred at Chelyabinsk and Tunguska, he also held a pertinent honor. In 2008, the first asteroid found in space prior to impacting Earth was discovered; Jenniskens and his team, carefully plotting out its terminal trajectory, then found forty-seven surviving fragments all over Sudan's Nubian Desert.[24] He arrived in Botswana almost immediately after the June 2018 event, nested himself within Gabadirwe's squad, and got to work.

Scout's orbital extrapolation helped, but the possible meteorite fall area was a 1,400-square-mile patch, about the size of Rhode Island. There is no special piece of technology that can scan a vast region and find meteorites; like detectives looking for pieces of evidence at the scene of a crime, all you can do is grab your team and slowly make your way through the grass, trees, and fields, your eyes glued to the ground. Hoping to narrow things down, the search party visited local businesses and gathered all the CCTV footage of the meteor's fireball they could. There wasn't much of it, but some security camera footage obtained from a hotel and several gas stations let them narrow down the meteorite fall location considerably.[25]

It turned out that if there were any shards from the stars, they had fallen in, of all places, not a wide plain, valley, or series of fields, but the Central Kalahari Game Reserve. The meteorite hunters all shared the same thought: you have got to be kidding me.

This national park, the size of the Netherlands and the second largest wildlife reserve on the planet,[26] is a labyrinth of life: trees and scrub bushes, dunes and grass; giraffes, gemsbok, wildebeest, bat-eared foxes, warthogs, and ostriches. It's breathtaking. But at that moment of revelation, all Gabadirwe and his cohort were thinking about were the many, many apex predators that also called the Kalahari home. Wondering if—perhaps hoping that—they had the wrong location, they checked in with some of the wildlife rangers, who confirmed that they saw the fiery plume enter the park.

There was one upside to the meteorites landing there. "We realized that some people might be after it," said Gabadirwe—meteorite poachers, essentially, hoping to make a profit. "But we knew it was in the middle of the game reserve," so that might put people off. Not Gabadirwe and his team, though. Left with no other choice, they surrounded themselves with an armed wildlife ranger escort and infiltrated the dangerous domain.

"It was a very . . . rare type of operation," Gabadirwe recalled. As they combed the grasses and sands, looking for the jet-black coating of meteorites—the glassy crust they often gain when burning up in the atmosphere—most of the herbivores simply sauntered on by. Sharp-toothed beasts, meanwhile, stared at the group of delicious humans for quite a while, occasionally getting distracted by a giraffe. "There were lions all over," he said. "It was not safe at all." Every now and then, the team would look up to see a leopard hanging in a tree, curious at their questing.[27]

Day after day, they came up empty-handed. And although the wildlife let their bipedal visitors get by in peace, the remnants of their dinners proved to be the biggest problem. Animal poop, it turns out, looks a lot like a meteorite until you poke it and your finger cracks open a crusty coprological doppelgänger.[28] This happened so many times that Gabadirwe lost count.

After a few weeks, the team was dejected. But Jenniskens refused to relent. "He was convinced something was on the ground," said Gabadirwe. He wasn't wrong. On June 23, the very last day of the hunt, someone took a walk from their new campsite and, just 300 feet away, found a meteorite, sitting ever-so-silently by a body of water. After confirming that it wasn't another find of animal poop, everyone erupted in celebration. The little shard, weighing less than an ounce, was named Motopi Pan, after the adjacent watering hole. Joined by several more searchers authorized by the powers-that-be, Jenniskens, Gabadirwe, and company eventually found twenty-three more meteorites, the charred remains of the 22-million-year-old Battle of Vesta.

Those fragments have since traveled the world. But some have their pride of place at the Botswana Geoscience Institute. It may not have the sway or funding as many major Western museums with populous meteorite collections, Gabadirwe told me. But, finally, it can get started. "It was such an amazing experience in my life," he said, beaming from ear to ear. "It was just *amazing*, honestly." A rare example of the heavens listening to the wishes of someone below.

Meteorites are chronicles of the earliest days of the solar system. They are inscribed with tales of epic acts of creation and destruction. And by studying pieces of various asteroids—the stony ones, the metallic ones, the ones full of powdery carbon, and all the in-betweeners—scientists can better understand how to disrupt or deflect their original city killer–sized hosts. By 2024, at least for astronomers, planetary scientists, and planetary defense researchers, meteorites have become ordinary to hold and behold. Humanity now owns a vertiginous mountain of them, over 50,000.[29] In this mix are some rare specimens, including meteorites from the Moon and Mars. The most scientifically remarkable meteorites are kept in special cases. Many are chopped up, forensically examined, and then stored in special science bags. You don't need prestigious credentials to experience this strange normalcy yourself: in the last few years, I've been invited to poke and prod—wearing proper gloves, of course, and in one case the full bunny suit with a face mask—plenty of meteorites myself, at various museums and universities around the world.

Going on the hunt for meteorites is always exciting, and in some cases, such as Gabadirwe's, it can be both perilous and exhilarating. Thanks to its great white emptiness, Antarctica is an especially good place to find blackened meteorites. Katherine Joy at the University of Manchester has been there plenty of times, evocatively describing it as an "expansive place where the sky and ice seem to go on forever."[30] At the end of the day, though, those crystallized teardrops from the dawn of time have become part of the job, another key to unlocking another of the cosmos's countless doors.

But what happens when a meteorite finds, well, the rest of us? What happens when the discoverer isn't a scientist, but someone who has barely given a second thought to the infinite space above? As the Wilcocks will attest, it can turn your life upside down.

It took a few days to be convinced, Rob Wilcock told me, sipping his coffee in a London bookstore's café. "I don't think about meteorites. I don't know how unusual they are." All he, his wife Cathryn, and his daughter Hannah knew was that they had about eleven ounces[31] of black sooty something "in a Waitrose freezer bag" sitting in their house. "In the utility room, of course," he added, next to a food processor. "Next door had twenty grams" (less than an ounce). Only after they had scooped up the mystery matter, and after having seen planetary scientists froth excitedly about a new meteor all over the news, did Rob come to a confident conclusion: that charred confetti in their house had to have come from outer space.

The Wilcocks hail from the English town of Winchcombe, which, until March 2021, was like several other settlements in the Cotswolds: bucolic, home to verdant countryside panoramas, masterfully manicured gardens, and a requisite castle. It was used to hosting visitors from around the UK and even farther afield. But on February 28, it received a visitor that had traveled a record-breaking distance. At about 10 p.m. local time, it exploded fairly quietly in the night sky, sending pieces in all directions. At the time, the only people that noticed were members of the UK Fireball Alliance,[32] a group of meteor aficionados led by the Natural History Museum in London. The alert went out: something just blew up over the county of Gloucestershire. What happened to it?

Like many residents of Winchcombe that serene Sunday night, Rob, Cathryn, and Hannah were chilling out at home. Hannah, who was upstairs in a road-facing bedroom, thought she heard a clattering noise,[33] a bit like a picture frame falling off the wall in another room. Unable to identify the source, she didn't think much of it until the next morning, when she looked outside at the driveway and saw a pile of Stygian soot. The trio went outside and stared at it. Did someone throw this at them? Did a cryptic beast leave it as some sort of message?

Rob texted his sons—POSSIBLY FROM SPACE?—accompanied by images of the peculiar pile. That's when one of his sons, Daniel, alerted them to fireball reports across the region. Rob leaped online, and he found an alert sent out by scientists to those in the county: if you live in the area and you have found something weird and rock-like that wasn't there before, for the love of God, please do not ignore it, or wash it away with a hose—keep it somewhere safe.

Imagine you're in this situation, and you aren't a scientist or someone who's made space their hobby. What would you do if you were confronted with a pile of extraterrestrial dandruff on your driveway? They didn't know if it was hazardous or volatile, Cathryn told me. Should they go near it? It looked like an anthill made of coffee grounds, and although it would have probably made an out-of-this-world espresso, they did what any sensible family would have done at that stage: put on rubber gloves, scoop it into polyethylene sandwich bags and plastic yogurt pots, brush in the smaller bits with toothbrushes and stainless steel knives, seal it all up, and put it in the house.

Fireballs had been detected over the UK before, but a meteorite hadn't been found from any of them for thirty long years. Desperate for news of any sightings, planetary scientists put out the call to the public for help. The Wilcocks uploaded the images of their tagged driveway to the network, where the meteoriticist Richard Greenwood, from the Open University, saw them and immediately knew they were genuine. On March 3, he visited the family, and was soon joined by the Natural History Museum's meteorite master, Ashley King. A careful examination of the samples in the family's garden made it clear that this was not only a meteorite, but a rare, carbon-filled,[34] and astonishingly primeval bit of debris from the solar system's conception.[35]

This tsunami of excitement and attention threw the family for a loop. "When it hit, we were unsure what to do," Cathryn told me. "TV crews were coming up and knocking on the door, looking under the car with cameras, asking us for interviews." It also happened during an intense COVID-19 era lockdown, and wariness about infection was high, so all their interactions with the scientists had to happen outside.

"Anybody walking past our house could see scientists just wandering about for a few days. Everybody in Winchcombe knew exactly what it was about," said Rob.

"Well, I'm not sure everybody knew," amended Cathryn. But those who wanted to find out where the pictures of their sighting came from went all-out to succeed. "One camera crew tracked where it was just by looking at the paving stones on the drive. It was just crazy."

They were advised by some of the scientists not to take on the legions of journalists swarming outside their house, and for a while, they turned them away. "As soon as you speak to the press, you'd be thrown to the wolves," Rob said, recalling the Natural History Museum's ominous advice.

"But in the end, we thought: this is daft. It's a good news story, let's share it with the world at large," said Cathryn. Hannah, who serendipitously worked with the BBC Science Unit, helped put them in touch with a cadre of friendly media members,[36] and their concerns melted away. It was intense, but "all the journalists were really lovely," Cathryn told me.

Why, I ask, were they initially worried about all the press attention that came along with such a dramatic scientific story? Rob said a major concern was of the criminal variety: an unscrupulous person might be bold enough to break into their house and steal it. "We got a sense from the scientists that it was valuable. And if we told the world that it's worth, I don't know, £100,000 or something, would that make us a target for thieves?"

But surrounded by a phalanx of scientists, press, and good-natured members of their community, they were shielded from burglars. They instead got to revel in a once-in-several-lifetimes spectacle. When Ashley King, standing on their patio and holding a large meteorite chunk, told them that it was 4.65 billion years old, their jaws fell earthward. "The hairs stand up on the back of your neck. It's scary! It was just amazing really," said Cathryn. They fondly recall scientists on their hands and knees, crawling around their property, extracting tiny meteorite particles from the ground with tweezers and toothbrushes.

Along with a small chunk that landed on their neighbor's property, a team of scientists combing nearby fields found additional meteorites hiding among sheep poop,[37] bringing the recovered mass of the Winchcombe meteor to 1.3 pounds,[38] no heavier than a basketball. The Wilcock family had taken delivery of just over half of the grand total.

By the time I met Rob and Cathryn in 2022, the couple had been transformed. "We've loved meeting so many scientists in the last year," Rob told me, reeling off various space-based factoids he had picked up since the discovery. They now had a new appreciation for "all the stuff out there that's millions of miles away, and [for] our own planet." Soon, scientists experimenting on their space rocks revealed that it contained water, bolstering the hypothesis that Earth got its oceans and seas not from comets, but from a bombardment of soggy asteroids.[39]

As I spoke to Rob and Cathryn, I found myself won over not only by their sense of wonder, but also by how philanthropically minded they were. "Very early on, we decided we weren't going to try and make money out of it. We wanted the scientists to have it," Rob told me. Earlier that year, a bean-size piece of the Winchcombe meteor not owned by the Wilcock family, just 0.3 percent of the total recovered material, was sold for $12,600, more than 120 times the value of its weight in gold.[40] Meteorites don't always sell that well; some get bought at double the asking price, whereas some, like one that hit a dog kennel in Costa Rica in 2019 (don't worry, the dog is fine), sold for ten times less than auctioneers predicted. But even at the lower end of the scale, you can sell meteorites the size of the Wilcocks' for tens of thousands or hundreds of thousands of dollars. But they decided that their meteorite had a nobler destiny. "We wanted to give it to science so people could learn from it, and so people could be inspired by it," said Rob.

For their efforts, a segment of their driveway was preserved and displayed at the Natural History Museum, along with a chunk of the meteorite itself.[41] That visitor to Winchcombe turned out to be the most important meteorite to have ever been found in the UK. The Wilcocks seemed content to be recognized for their supporting role in its grand solar system saga.

Not good enough, I thought—before they left, they deserved a little something extra for all their humility and good will. So I let them in on a little secret. I filled them in on DART, the global effort to find asteroids before they find us, and the fact that we can't know how to deflect or destroy city killers, country crushers, and planet destroyers unless we know the constitution of our targets. The Winchcombe meteorites came from an especially rare type of cosmic rock. You've not just contributed to our comprehension of the solar system, I told them; you've also aided the effort to save the world. They looked at each other, then back at me. Rob shook his head and smiled. "We had no idea about that aspect of it at all," said Cathryn.

Meteorites are instrumental to our global, asteroid-killing efforts. We're lucky that people like the Wilcocks exist to bolster that campaign. But scientists are also impatient, and ambitious. They don't want to wait for the stars to send them a clue. Ideally, they want to fly up there and encounter the rocky antagonists themselves. Earth's planetary defense system may live or die on the success or failure of DART. But the road to September 26, 2022—the day of impact—had already been paved by some truly daring space missions, each visiting a comet or asteroid to leave a history-making mark of their own.

V

Never Tell Me the Odds

DART is more than just a planetary defense mission. It's a technology demonstration, not much different from when companies debut a new piece of equipment at a trade show and show it off to curious customers and investors. Except this time, the demonstration is taking place millions of miles from home—and it's showcasing how this technology can save millions from potentially killer asteroids. Apart from proving the overall viability of a deflection mission, the technological prototypes attached to DART get their moment in the limelight to show everyone—including future spacecraft builders—that they can do what they do with pizzazz.

The NEXT-C thruster was one such gizmo, an experimental booster that wasn't critical to the spacecraft's journey, but one that might be able to give DART a helping hand while showing off its capabilities. It was switched on not long into the flight, on December 18, 2021—and after blasting out some ions into deep space, it was switched off again. Everything seemed absolutely fine.[1] But when Elena Adams and her team checked the dashboard stats the next day, it was clear that everything was, potentially, terrible. "Oh, wait a second, our voltage spiked," she recalled saying at the time. "It damaged our spacecraft a little bit.

It caused current spikes. It pumped some of the current into the spacecraft body. That wasn't expected."

Unexpected moments that happen midflight during a mission like DART are not like the surprising things that happen to most of us on a daily basis. This wasn't like being fouled on by a bird on the way to work, bumping into an ex, or realizing you put on your shirt backward. What happened to DART was more like a seagull dropping a rock onto your head with a not-insignificant chance of killing you. The spacecraft had been accidentally supercharged, and it wasn't immediately clear why. Adams wondered, "Oh. Is it going to die? Because that would be really bad."

Action stations. While checking on the health of the spacecraft's other instruments, they worked the problem. After noticing that the spike coincided with the thruster's activation, they ran the same NEXT-C test on an engineering model of the thruster they had built and kept back on Earth. Zap. Electrical surge detected. Okay everybody: nobody switches NEXT-C on again. Ever.

That was too close. To everyone's enormous relief, DART appeared copacetic; it was still heading toward its target. Nobody knew for certain what was going to happen when it reached Dimorphos—it might fail to deflect it, fudge it completely and miss, or hit it too hard and smash it into pieces. But it had to reach the target. That was the lowest threshold, the bare minimum. Its human copilots, like Adams, were fiercely determined to make sure the spacecraft accomplished at least that much. Otherwise, DART's robotic predecessors would never let them live it down.

Comet Tempel 1 never knew what hit it.

Barely four miles long, this diminutive ball of ice had been zipping around the solar system for millions of years, minding its own business. It originally came from the frigid realm known as the Oort cloud, but at some point during its life it decided to go off on an adventure. It said its fond farewells to its family and migrated toward the inner solar system, orbiting the Sun once every 5.6 years[2] and going no farther out than Jupiter.[3] And of all the worlds it flew by, its favorite was a blue-green orb decorated with a silver moon.

Tempel 1 was named after Wilhelm Tempel, a telescope wielder who first spotted it outgassing in 1867. The comet didn't mind having a human moniker. Humans were harmless, curious critters: they loved making things and asking questions. Perhaps, the comet thought, they would pop over and say hello one of these days.

Little did it know that, in January 2005, a spacecraft pieced together by some especially intrepid humans had just been launched into space to do exactly that. By June, it had traveled 268 million miles and had managed to catch up to the comet.[4] *Oh, hey there! Were you sent by the Earthlings to say hello? How thrilling!*

The spacecraft said nothing.

Perhaps it was shy. Comets, at least to humans, are enormous entities. Tempel 1 threw out a bit of friendly vaporized ice as a welcoming gesture, and let the metal box have all the time it needed to extend its greetings. As fate would have it, this NASA spacecraft was all set to meet Tempel 1 on July 4. Its mission had nothing to do with Independence Day, but it was about to make a lot of fireworks.

The previous day, an 820-pound copper object had silently emerged from the spacecraft. At first, things looked peaceful: 625,000 names adorned the copper envoy, and the cameras on the main spacecraft were filming the whole thing.[5] *It's coming to say hello on behalf of hundreds of thousands of people, how lovely!* And at the speed it was going, it looked like humanity was keen to say hello as quickly as possible—like, really keen.

By the time July 4 arrived, the copper messenger had reached a relative velocity of 23,000 miles per hour. *Whoa, slow down there, buddy! I'm not going anywhere. No need to rush the landing, you know?*

It did not slow down.

Hey, okay now. I mean . . . I guess the humans know what they're doing. I'm sure this is perfectly normal. The comet, although anxious, extended a vaporous hand to the incoming ambassador from planet Earth. *Hi there! I'm from the outer solar system. I've been admiring your planet for some time now, it's so great to meet—*

Boom. An explosion equivalent to 4.7 tons of TNT shook the comet as the copper object punctured its face. Everything was bathed in a white light, one that became so luminescent it almost blinded the mechanical eyes of the spacecraft orbiting nearby. Rays of dust and ice shot out in all directions, screaming into space. And back on Earth, everyone at NASA erupted into teary-eyed cheers.

The success of the mission, named Deep Impact, was far from guaranteed. Nobody had encountered a comet quite like this before. Not long before impact, Charles Elachi, the head of NASA's Jet Propulsion Laboratory, said that this mission was like hitting "a bullet with another bullet while watching from a third bullet."[6] But they had done it: they had caught up to Tempel 1, dodged potentially dangerous outbursts of gas, and shot a comet with a mostly self-guided copper projectile. The comet's insides were pouring out into space from a 500-foot wound, and for the first time in human history, scientists could see what a comet was made of. It's a good thing that comets aren't really sentient.

Lindley Johnson, the soon-to-be head of the Planetary Defense Coordination Office, was thrilled. "We had these artists' depictions about what we thought would occur when that impactor hit that comet and blasted out this ejecta cone around it. I looked at those depictions and I said, ah, no, it's not gonna look like that," he told me. "And I'll be darned that when it actually hit and we saw those images come back, it bore a striking resemblance."

The visuals were so breathtaking—a vast ghost throwing out trillions of shimmering jewels in all directions—that Kelly Fast, Johnson's soon-to-be second-in-command, took her elementary school–aged kids to see stills and footage of the impact in the middle of the night. Megan Bruck Syal, the nuclear weapon (or NED) planetary defense researcher at Lawrence Livermore, was an undergraduate at the time. She could have taken her nascent career in any direction, but when a scientist on the Deep Impact team went to her college and showed her the footage, her destiny as a planetary defender was set in stone.

Everyone, from scientists to the press, wanted a piece of the cinematic action. "We broke the internet back then," Jessica Sunshine, a comet and asteroid expert at the University of Maryland, told me. Now part of the science team for the DART mission, she was once a co-investigator on Deep Impact. Yes, uncovering the history of the solar system by excavating the insides of a primordial comet mattered. But what mattered most to Sunshine was that this was accomplished by shooting something at it.

"To me, scientifically anyway, I thought: this is what we should be doing. We're experimentalists. Why do we think of missions as entirely passive?" Using impacts is a maximalist method of learning about the cosmos. "You could shoot Mars, you know? There's no place we couldn't learn something. We could stop debating how much subsurface ice there is."

The difference between Deep Impact and DART is that the former was a science mission, not a planetary defense mission. Its objectives were chiefly academic. That copper bullet certainly gave Tempel 1 a nudge, but thanks to the comet's mass, it was insignificant. But that a spacecraft semiautonomously piloted part of itself into a target zipping through space—perhaps the sort of object that will one day find itself on a collision course with Earth—meant that, unofficially, this was a planetary defense test of sorts.

"We knew we were doing planetary defense. We just weren't allowed to say it," said Sunshine. A planetary defense office didn't yet exist, and the Deep Impact scientists were told to publicly state that it was a science mission. "NASA wasn't ready for it, for sure, to talk about spending money on planetary defense. That was not in their bubble of things."

DART wouldn't exist if not for Deep Impact and its resounding success. "The reason I'm on the DART team is because I'm sort of the living legacy of Deep Impact," said Sunshine. And it wasn't the only mission that paved the way for DART. Several others followed Deep Impact's lead, leaving behind more technological bedrock—while also handing the community a distressing scientific puzzle. It turns out that comets and asteroids do not do what you think they will do when you hit them. And that would one day give DART, and planetary defenders, an enormous headache.

Comet Tempel 1 may have been left with a terrible impression of humans, but just a few years later, Comet 67P/Churyumov-Gerasimenko—67P for short—had a considerably better experience. This time around, the European Space Agency genuinely wanted to say hello, and calmly study the comet's surface, not whack a great big hole in the side of it.

Originally discovered by Soviet astronomers in 1969, Comet 67P had, for most of its life, traveled no nearer to the Sun than just inside Jupiter's orbit. But astronomers think that in 1840 it got a little too close to that gas giant. The planet's immense gravitational pull shunted the ice ball onto another highway, putting it on a trajectory that regularly takes it quite close—but harmlessly so—to Earth, giving astronomers a hell of a light show when its ices periodically and explosively transform into ethers.[7] As was revealed long ago by the studious gaze of specialized telescopes, a lot of Comet 67P's ice is water—but, some wondered, is it the same water we have on Earth? Curious to know if comets were the primary source of our rivers, lakes, and oceans, the European Space Agency decided to chuck a spacecraft its way and go full Poirot.

That mission was named Rosetta, after the ancient stone tablet found in 1799 during Napoleon's military campaign in Egypt; inscribed with two forms of Ancient Egyptian, and Ancient Greek, its discovery gave researchers the ability to finally translate those mysterious hieroglyphs. In much the same manner, Rosetta would decode the nature of cometary water, allowing scientists to compare it to our own.[8]

And so, on March 2, 2004, it was flung skyward atop an Ariane-5 G+ rocket, leaving behind the humid rainforests of the European spaceport in Kourou, French Guiana. A decade later, after a trajectory-altering galivant around the inner solar system, Rosetta came face-to-face with Comet 67P—which, incidentally, resembled an iced-over rubber duck—on August 6, 2014. For a few months, the spacecraft hung about at a comfortable distance behind its new pal, taking reams of photographs.[9]

Rosetta revealed that comets were not smooth orbs of ice and nothingness. They had topography: mountains, craters—so many craters—volcanic-like jets of icy matter sibilating and fountaining into the dark. Dust and snow bounced about, like weather capable of defying gravity. There were cliffs and valleys. There were ripples,[10] rhythmic features that resembled sand dunes. And its tail of crystalline ices and vapor looked positively mythical, a gleaming cape of infinite forms. It was bewitching to behold.

By September, Rosetta had approached the comet and instead of chasing it began to circumnavigate it, officially becoming the first spacecraft to orbit a cometary nucleus.[11] Then, on November 12 of that year, Rosetta played the ace up its sleeve: Philae—named after an obelisk that played a similar linguistic role to the Rosetta stone[12]—left its taxi of ten years and began to drift toward the comet. The European Space Agency was going to try something nobody had dared attempt before: Philae was going to touch down on a comet and stay there, conducting as much scientific sleuthing as it could before its solar-powered batteries spent too long in the shadows and ran out of juice.

Things were already off to a precarious start. The lander spacecraft was relying on its thrusters to not only swim toward its parking spot but also to stabilize its landing. A crash-landing would be awful, but it would be just as grim if Philae bounced off the surface. About 2.5 miles wide, Comet 67P would be sufficient to kill billions of people if it ever found its way to Earth (fortunately, astronomers know that this isn't in the realm of possibility). But it is still small for a comet, meaning it has very little gravity. The lander had to touch down on its surface at just the right speed, or else it would ricochet off into deep space with no chance of recovery. And just before it left Rosetta, scientists found that its thruster system wasn't working properly.

What else could they do? This was the only chance they would get to touch down on a comet. Luckily, they had another grappling mechanism built into the lander that still seemed operational: a pair of harpoons that would fire into the ice and, all going well, hold the trampolining robot in place. The command was sent, and the lander began its seven-hour-long flight toward the frozen world. At the end of that timer, the ground team knew that Philae had to have made contact, but the distance from Earth meant that they had to wait twenty-eight agonizing minutes to receive what they hoped would be the touchdown signal from the spacecraft.[13]

They got it. But something was wrong. The dashboard seemed to show three touchdown signals—two separated by two hours of time, and a third coming just seven minutes later. What . . . the heck? That didn't seem great. Soon, the truth became clear: the lander did ping off the comet but somehow managed to head back toward it; it bounced up again, before finally flumping down onto the surface and staying put. Like its thruster system, Philae's harpoons malfunctioned: they failed to properly fire, and yet the dishwasher-size spacecraft still somehow managed to stick the landing.[14] Even specialized screws attached to its base couldn't pummel into the ultra-rigid ice below, so Philae was holding on to Comet 67P with nothing more than luck and magic.

Sensing its time may rapidly run out, it got to work almost immediately, sniffing the icy strangeness around it, and sending all the data it could back to Earth. Perched up against an overhanging cliff, its solar panels were receiving little sunlight, and the battery died just fifty-seven hours after it had landed. But it was enough. Between Rosetta and Philae, scientists determined two key things: this comet possessed the kinds of chemical compounds that may very well have been vital to the first living things—but its water bore a signature different to that of Earth's. Comets may have delivered some water to our world, but perhaps another source was the primary philanthropist.[15] Some suspected asteroids. Remember the meteorite that fell onto the driveway of a stunned family in the middle of England in 2021? As it turned out, that meteorite contained water with a near-perfect match to Earth's.[16]

It's a beautiful story, with a sentimental ending. Philae emerged from its slumber on June 13, 2015, and intermittently spoke to Earth over the next few weeks—a short reactivation, spurred by sunlight falling onto its solar panels. But that was it for the lonely robot: it was too shadowed, and too damaged, to escape its eternal dream. The next year, Rosetta managed to spy its expired partner, concealed in a benighted gulch.[17] Soon after, Rosetta was commanded to plummet toward the comet, soaking up its surroundings as it went, right up to the moment it crashed and its signal flatlined. Back home, in the European Space Operations Centre in Darmstadt, Germany, its creators cried, laughed, applauded, high-fived, and hugged one another. On September 30, 2016, the mission met its end.[18]

While many were occupied by the watery composition of Comet 67P, a few scientists remained fixated on the bounce. The gravity of the comet was low, sure, but Philae bounced twice. Was something else afoot here? After spooling through Rosetta's imagery, the team studied the imprints the lander left behind on its bouncy final voyage.[19] And in 2020, just before Halloween, they announced something appropriately spooky: the lander failed to puncture the ice, but it may not have been because the ice was rock solid; instead, the ice was almost foamy, not dissimilar from a cappuccino's crown. It was softer than the lightest dusting of snow.[20]

How could that be possible? I called Michael Küppers, one of that study's authors and a member of the Rosetta mission, to ask. "For some time, we were really puzzled, and we didn't know what to say," he told me. "It was extremely soft. Half the density of water." If you were to walk on the surface of Comet 67P, it would be like stepping on a platform made of endless bubbles atop a bath. This meant that the comet was barely holding itself together by its own gravity—and yet, there it was, barreling through space.

Planetary defense had long been based on the notion that NEOs were firm objects, the sort that would respond in a straightforward manner to anything trying to hit them. But impacting a frothy, fragile entity in the hope of deflecting it was a far more complex endeavor. It wouldn't absorb the energy of the spacecraft in a linear way. It may be deflected, but not how everyone intended. Worse, it may be disrupted, breaking it into pieces that could still head toward Earth.

The Rosetta mission revealed that deflecting a comet would be tougher than anyone previously thought. Michele Bannister, the DART team member in New Zealand, put it best. "How do you push something like that? How do you fight with foam on a beach?"[21]

Good question. The upside is that compared to asteroids, comets are considerably less likely to hit Earth: they mostly linger in the outer solar system, with just a handful daring to pay a visit to the inner solar system. The chances that we will one day be forced to reckon with a cometary impact are not zero, but for the next century at least, those odds are very close to zero.

The downside is that asteroids are just as strange as their icy cousins. They can be lumps of metal, piles of boulders flying in formation, or as fluffy as comets. All these possibilities, and everything in-between, exist, and each would respond differently to a kinetic impactor like

DART or a nuclear device rushing to meet them. And that unnerving reality was made abundantly clear by a truly bonkers space mission—one that pulled off a deep-space heist, stealing priceless matter from the dawn of time and, miraculously, getting away with it.

Throughout our conversation, I failed to understand how Yuichi Tsuda remained so understated. He was, after all, the greatest thief in the solar system. He sent a "Peregrine falcon" millions of miles into space to break into a rocky vault 4.6 billion years old, got the treasure within, and snuck off back to Earth. And yet, somehow, he was the one grateful to me for simply chatting to him about it. "Thank you for your interest," he said, sincerely.

Hayabusa2—the sequel to a similar mission, both named after the speedy Japanese Peregrine falcon—was a true masterpiece of engineering. Many things had to go right—in fact, everything did, because if one maneuver failed, or one key instrument did not do its job properly, all would have been lost. "It was very challenging," said Tsuda. "There were many hardships to realize the beautiful touchdown." But they did it. They robbed an asteroid, and in the process found out that asteroids are extremely odd.

Tsuda, the project manager of the mission at the Japan Aerospace Exploration Agency, or JAXA, did not set out to be an off-world pirate. "My interest in space starts very early, when I was six or seven years old. I was not so interested in space itself at that time. I got interested in making something, or crafting." He loved aircraft, spacecraft, ships, trains, and all the work needed to piece them together. It makes sense that one day he would upgrade to space, culminating in the creation of Hayabusa2, a burglar that hoped to pickpocket a 3,000-foot-long, diamond-shaped asteroid named Ryugu in dramatic fashion.[22]

Ryugu—whose awesome name derives from the palace of a dragon monarch from a Japanese folk story—hangs about between Earth and Mars, making it close to home but of no danger to us. A so-called carbonaceous asteroid, it is one of the most commonplace,[23] but that is precisely why it's valuable to scientists: these asteroids would have played a constitutional role in the world-building phase of the embryonic solar system. Knowing what they are made of, then, is crucial to understanding our own existence. To bolster this quest, Hayabusa2 was going to land on Ryugu's surface, blast out a small crater, and scoop up the debris before bringing it back home.

This was primarily a science mission. But as with Deep Impact and Rosetta, every visit to an asteroid has planetary defense implications. Ryugu may not be the type of asteroid the community is chiefly concerned by—it's pretty big and, like most of its similarly sized siblings, it's easy to spot—but the action of shooting at it would have a comparable value to the Deep Impact mission: How would such an object respond to being nudged? The answer was essentially unknowable when it came to Ryugu, because this dragon's palace was anything but monolithic. It is a rubble pile asteroid, a bunch of rocks barely bound together by their own gravity.[24] Hit one of those a bit too hard and you've got yourself not one, but dozens of asteroids.

The Hayabusa2 team had to be ultracareful. After a three and a half years voyaging through space, it caught up to Ryugu in June 2018. Unhelpfully, the asteroid was spinning like a dreidel, making any touchdown attempts especially tricky. But, at first, everything seemed to go perfectly: between September and October, two rovers and a lander left the falcon and parked on the asteroid, conducting an intensive scientific investigation shortly afterward. The falcon itself dove close to the surface before pulling back again in November—a trial run for the real deal.[25]

On February 21, 2019, the first of Hayabusa2's opuses unfolded. It slowly approached the asteroid until it just about contacted the boulder-strewn surface, then, boom: it fired a tantalum bullet, sending ancient world-making matter scattering in all directions—some of which found its way into a little cache on the robot's shell.[26] Tsuda and his team, checking all the falcon's vital signs, quickly realized they had succeeded. That's miracle number one in the can.

This was but an amuse-bouche for what was to come on April 4. This time, instead of a tiny bullet, the falcon primed a copper cannonball armed with a bunch of explosives. All looked good, and Tsuda et al. gave the order to fire. A pause, as the message traveled across space to reach Hayabusa2, then, bang: the explosives were lit, the projectile emerged from the spacecraft at 4,500 miles per hour—about three times faster than a fighter jet—and splashed into Ryugu.[27] Blackened fragments leaped silently into space, and the rubble pile rippled and trembled, almost as if it were alive.

The idea was to create a thirty-foot-deep crater to expose the subsurface material that had not been altered by eons of starlight and cosmic rays.[28] And on July 10, those samples were collected when the falcon gently prodded a spot close to the crater—as it turned out, not in the crater itself, as it was deemed too difficult to safely land inside it—and, with the aid of another small tantalum bullet, blasted that pristine rocky matter off the surface before quickly scooping some of it up.[29] Miracle number two had been achieved. *Top Gun*'s Maverick has got nothing on JAXA's team.

Scientifically, this mission was an unbelievable success. Hayabusa2 returned to Earth in December 2020, dropping the asteroid dust capsule into the atmosphere; it landed safely in the Australian outback, and was picked up by a recovery team, who hoped that the capsule had not been damaged and exposed to the elements upon impact.[30] It was found intact. Inside was "just a tablespoon" of extraterrestrial material from the genesis of the solar system, said Tsuda—fifty times what they had hoped to steal from Ryugu.

In 2005, JAXA's original Hayabusa mission had attempted to gather samples from the asteroid Itokawa—and it did, but a series of technical glitches and errors ultimately meant that a vanishingly small amount of matter was transported to Earth.[31] It was nonetheless a record-making mission: the first spacecraft to safely touch down onto and lift off from an asteroid, and the first to return surface samples back home. But Hayabusa2 showed everyone how it was supposed to be done.

Were Tsuda and his team confident they would succeed? "There's always a small risk of failure. We have to always accept that," he said. In this case, that risk was less than 1 percent—highly confident, I would say, considering, in Tsuda's words, "this was the first time in human history we've tried this." The engineering team had a sort of mantra: "To have fear in the right way is important for success." No matter how thrilling a space mission is to work on, the prospect of failure can be poison to the minds of those managing it. How, I wondered, did Tsuda handle his own mission? Was he always on edge, or did excitement regularly subsume his anxiety?

"It's a difficult question," he told me. After each stage of success, "we celebrated a lot—with beer, or with wine." He paused. "It's very important to manage a team, to put the right milestones . . . small enough so that the team members can feel the success, not ten years later," he said. "We had a lot of reasons to celebrate. We made full use of that to motivate ourselves, and to make ourselves relax."

Tsuda's team had to be extremely careful as they planned the details of the mission. Interacting with a rubble pile in space was difficult to envisage on Earth, though they gave it their best shot. At a Tokyo testing laboratory, they placed various materials that they thought might mimic Ryugu, from concrete to rocks to metals to sands, at the business end of a gas-filled hypervelocity gun and fired round after round. When all the dust had finally cleared, they guessed that any crater they would make on the real deal would be small, just a few feet across. Yet the actual crater turned out to be bigger by an order of magnitude.

Ryugu was known to be weakly held together. But their falcon showed that it was bound by "almost nothing," said Tsuda. The huge chasm they left behind, and the huge volume of matter that escaped into space, showed that even modestly sized asteroids are improbably weak—and, therefore, vulnerable to being accidentally disrupted rather than deflected.

With all this in mind, I began to wonder why JAXA was not a major partner in a DART-like mission—or why they had not beaten America and Europe to the punch, considering how good they were at such swashbuckling space missions. It turns out that Tsuda originally wanted Hayabusa2 to be a planetary defense mission, but the nation's post–World War II legacy denied him that chance.

Japan's postwar constitution, overseen by the American-led Allied occupation,[32] was a successful attempt to turn an absolute monarchy into a thriving and peaceful democracy. The text's Article 9 states that war will be forever renounced, and as a result Japan is only permitted to have a restricted and largely home-based military known as the Japan Self-Defense Forces.[33] Offensive military operations abroad were and still are banned, even though a law allowing certain overseas operations was passed in 2015.[34] Consequently, any changes to the nation's military capabilities are fraught with controversy, to the point that any technology deployed in the use of defense is seen as concerning—even if it's to defend the entire planet from a hazardous asteroid.

"Our original plan, actually, was just like the DART mission," said Tsuda. "If we were successful, we could have had a DART-like mission a few years ago. Our first proposal was to make the impactor spacecraft fly in parallel with the Hayabusa2 main spacecraft. And after the arrival of the main spacecraft to Ryugu, the impactor spacecraft was to hit Ryugu." Initially, Hayabusa2 was a bit more like the European Space Agency's deflection mission to Didymos, Hera—or AIM, the original proposal, which would have gotten there before DART made its mark. "It was like a DART-plus-Hera mission," he said—JAXA would have also gotten their Hera-like craft to retrieve the impacted material and return it to Earth. "But it was not authorized. Defense technology makes us a bit nervous." The mission "was thought at that time to be a little bit dangerous."

There is a world in which JAXA would be the lead agency in protecting the planet from dangerous asteroids. Did Tsuda regret that we are not living in that timeline? He seemed a little melancholy while talking about that alternate reality, but Hayabusa2 is widely recognized as one of the most game-changing space missions in human history, an accolade that would make any spacecraft manager proud.

His falcon still has one last adventure left to go. "We are going to visit two more asteroids," said Tsuda. The extended mission is dubbed Hayabusa2# (pronounced Hayabusa2 Sharp, like the musical notation), to describe the uptick in the difficulty of the mission's final objectives, while also being an acronym for *sharp*: Small Hazardous Asteroid Reconnaissance Probe. The spacecraft will fly close to one small asteroid in the next few years, before reaching a second, a rock not much larger than the Chelyabinsk meteor, in 2031. "We still have one projectile left on the spacecraft." They won't be able to bring any more samples back to Earth, but they can still attack. After all, "the best way to know [an asteroid] is to shoot at it."

Hayabusa2 was sent into space in 2014. Two years later, the United States was hot on its heels with the launch of its own van-size thief. It's mission: become the first American spacecraft to steal a sample from an asteroid. It's target: Bennu, a near-Earth asteroid 1,600 feet long named after an Egyptian deity.[35] The (full) name of the spacecraft: Origins, Spectral Interpretation, Resource Identification, and Security-Regolith Explorer—better known as OSIRIS-REx.[36]

Dante Lauretta is America's Yuichi Tsuda. He is not quite as understated as Tsuda. I can easily imagine him wearing aviators, regardless of how sunny it may be. I had spoken to him a few times before, and he had always come across as jocular, but nothing gets him quite as animated as OSIRIS-REx—because, like Hayabusa2, everything about it is snazzy, including its backronym. I wonder who came up with that?

"That was me. That was the first thing I did after I met with Mike and accepted the job," he told me, with a grin. "Osiris has that dual nature. Osiris is the god of agriculture. The god of life and growth. But he also died a horrible death. He was murdered by his brother and became the god of the afterworld. So he represents that dual nature of life and death. And asteroids represent that as well": they can kill billions, but also bring water and organic chemistry to worlds that may accommodate life. Okay, very neat. But why REx, as in "king," for the Regolith Explorer part of the name? That came about after the mission concept was given a higher spending cap during a redesign. "And it sounded like a dinosaur, so that was cool."

Lauretta wasn't supposed to oversee OSIRIS-REx. The Mike he referred to was Michael Drake, the director of the University of Arizona's Lunar and Planetary Laboratory. Along with various partners, the two of them conjured up the initial design back in 2004 and shepherded the mission to its official acceptance by NASA in May 2011. "Mike passed away in September, four months after we were selected. That's when I became the principal investigator."

OSIRIS-REx—built and operated by Lockheed Martin, the US aerospace and defense company—was equipped with a different suite of scientific instrumentation than Hayabusa2, as well as a different manner of sample collection: it would make use of a ten-foot-long arm in a pogo stick–like method.[37] The plan was to find a boulder-less spot to poke, something known as the Touch-And-Go, or TAG, event. Upon making contact, a pressurized nitrogen bottle at the end of its arm would force gas into the asteroid and send pebbles flying,[38] some of which would hopefully enter a collection chamber before the spacecraft ascended back into space.[39]

OSIRIS-REx arrived at Bennu in December 2018, and spent the next twenty-two months carefully observing and mapping the asteroid's surface and trajectory around the Sun. Then, on October 20, 2020, its do-or-die moment arrived: a site named Nightingale,[40] which looked to be part of a crater filled with freshly exposed subsurface material, was where the TAG was to be performed.

Getting to this point had been a bit of a nightmare. Like the DART mission during its construction, the pandemic had mercilessly hit the flight phase of OSIRIS-REx. "It was in the middle of October 2020, in the middle of COVID with no vaccines or anything, so that was just horrible. And the team couldn't really get together," recalled Lauretta. Masked up and waiting nervously for an effective vaccine, his team did what they could to work around the omnipresent virus without compromising their health.

Bennu also presented its own obstacles. Though the asteroid's surface was initially thought to be something closer to a sandy beach, the spacecraft revealed a world covered in huge rocks and boulders, the sort that would make a safe-but-brief touchdown on the asteroid, and a successful sample collection, deeply uncertain prospects.[41]

They had no choice but to take the plunge. On October 20, OSIRIS-REx descended upon Nightingale. Everything about the downward journey, guided by the spacecraft's autonomous navigation software, was nominal. But the moment the craft made contact with Bennu, the asteroid sprung its trap. "We just sunk right in, and all the material goes flying everywhere. That was a big surprise," recounted Lauretta. Instead of delivering a light prod, the spacecraft was partly swallowed by the amorphous ground beneath it. "We thought for sure there would be some strength to that surface. How can it hold up that giant boulder and not hold up the spacecraft? It was like quicksand, it just kept on going." They knew it was a rubble pile, but this was like punching a pit of plastic balls—the entire asteroid was held together by the most tenuous of threads.[42]

There was nothing they could do back on Earth. If the craft sank any farther, it would be game over. Fortunately, OSIRIS-REx was a smart cookie: it decided to fire its thrusters just in time, propelling itself out of the asteroid's maw and into the safety of space. "I think if [the spacecraft] hadn't fired [its] back-away thrusters, we would have lost the spacecraft. It would have disappeared into the asteroid," said Lauretta.

Not only did OSIRIS-REx escape, but it gathered so much material that the door to the sample chamber was jammed open for a few days, allowing the hard-gained treasure to leak into space before it was eventually closed.[43] Then, after a nearly three-year journey back through the solar system, the spacecraft flew by Earth. On September 24, 2023 the craft dropped off its sample capsule—one that safely landed at an active military test range in Utah, at 8:52 a.m. local time, to cheers, applause, tears, and palpable relief. I originally pictured an aviator-glasses adorned Lauretta eagerly leaping out of a helicopter to grab the capsule, before holding it up like a well-deserved alien trophy. But the first to walk up to the capsule was, surreally, an explosives ordinance expert who had to make sure that there were no unexploded bombs, mines, or shells at the site before Lauretta was permitted to approach.

The parallels between Hayabusa2 and OSIRIS-REx are explicit. Even Bennu and Ryugu have their similarities: both are close to home, both are full of carbonaceous material, and both are loosely bound rubble piles. But the day of the TAG maneuver reminded Lauretta of the upcoming DART mission—specifically, Dimorphos, DART's target. "Dimorphos looks a heck of a lot like Bennu," he said. It's not a single stone, but a collection of rocks. "It's like a liquid droplet. It's not a solid shard of material." How that smaller asteroid would respond to something hitting it extremely hard was anyone's guess.

The S in OSIRIS-REx stands for Security. It refers to our planet's security, which means that unlike every other mission before it, OSIRIS-REx was also recognized as a planetary defense mission. That, in part, is because Bennu—unlike Ryugu, Comet 67P, or Comet Tempel 1—isn't harmless. Astronomers had previously calculated that Bennu stood a one-in-2,700 chance of impacting Earth in the next couple of centuries: low, but not sufficiently low to be dismissed. Part of OSIRIS-REx's mission was to follow the asteroid and chart out its trajectory to an unprecedentedly accurate degree. It did just that, and found that between now and the year 2300, there is actually a one-in-1,750 chance of impact, a little higher than previously thought.[44] At the time of writing, Bennu's best shot at reaching Earth will come on September 24, 2182.[45] Its chance of success is still vanishingly small. But those odds mean that, for now, "Bennu is the most potentially hazardous asteroid in the solar system," said Lauretta. It's worth keeping a close eye on, for our descendants' sakes.

OSIRIS-REx did at Bennu what the aborted AIM spacecraft was intended to do for DART: arrive early and properly map out the asteroid, while providing the most accurate trajectory data possible. One day, without question, the world will need to deflect or disrupt an Earthbound asteroid. If our eyes on the skies are sharp enough, and the threat is found many years or decades before impact, then a scouting mission like OSIRIS-REx or AIM would ideally be sent ahead of the kinetic impactor or nuclear device. Forewarned is forearmed, and a successful defensive campaign may rest on the data that proactive reconnaissance provides, from the mechanical strength of the asteroid to its precisely plotted path around the Sun.

Such reconnaissance is so vital that it could be considered the third prong in the planetary defense trident, alongside detection and defense. As for the OSIRIS-REx spacecraft, it may have made its multibillion-mile asteroid delivery to Earth, but its adventures are far from over. Minutes after the sample capsule was released, OSIRIS-REx fired its thrusters and retreated from our planet's doorstep. It will soon slingshot around the Sun and eventually catch up with Apophis, a notorious asteroid that will zip by Earth on April 13, 2029. On that day, this 1,100-foot-long rock will get closer to the ground than plenty of our satellites.[46] That may sound scary, and for a time, it was. Shortly after it was discovered in 2004, the odds of it hitting Earth in 2029 were estimated to be one-in-37—unsettlingly high for an asteroid whose not-too-distant impact could wreak havoc on a continental scale.[47] That its namesake is the ancient Egyptian god that embodied chaos and darkness—one intentionally chosen by its discoverers—likely didn't help engender a sense of calm.

Fortunately, follow-up observations quickly reduced those odds—and, over the years, the asteroid's close flybys of Earth have allowed astronomers to track its movements so precisely that all possible impact dates within the next century have since been ruled out.[48] Nevertheless, OSIRIS-REx—or, rather, its extended mission, OSIRIS-APEX (Apophis Explorer)—will conduct a scientific drive-by around the time the asteroid passes under many of our satellites. This will give planetary defenders a good look at an S-type asteroid, a denser, stonier, potentially more cohesive object that also has plenty of siblings flying around the inner solar system.

Lauretta, though, won't oversee this extended mission. He isn't primarily a planetary defender, despite being a player in the global effort. A cosmochemist by training, he is focused less on preserving life than working out where it all came from. "I'm all in on: How did life originate on Earth? I've made it my primary intellectual focus," he said. "Can I move the needle at all on that question?" In 2022, all he could think about were the pieces of Bennu his team had stolen from the stars. He couldn't wait to "pick apart this sample, atom by atom, and figure out the history of the solar system."

Planetary defense? That's for the DART team to handle. And in early 2022, it looked like nothing would stop that most dramatic of space missions, not even that blasted NEXT-C experimental thruster and the spacecraft's accidental electrocution. Elena Adams and her team were, for a moment, all smiles. Then April arrived, and the star tracker—DART's second eye, the one used to determine where exactly it was in space—started to move. Something was nudging it, physically pushing it out of the way. And if they couldn't find out why, and they couldn't stop it, there was a chance DART would never find its way to Dimorphos.

VI

This Is How We Lose

Ludovic Ferrière was flanked by armed guards. It was his fourth day of captivity, and he had been escorted into an elevator inside the Ministry of Foreign Affairs. It was small, cramped. The doors closed. A jolt. They began to ascend. Ground floor. First floor. Then—another jolt. The elevator suddenly stopped, and the light went out.

This was it. This was the only chance he was going to get.

Almost without thinking, driven purely by the instinct to survive, he grabbed the rifle held by the guard to his left and pushed back into him, causing the guard to loosen his grip for just a second; the rifle fell to the floor with a clatter, and Ferrière kicked it away. Before the second guard could react, a pencil emerged from Ferrière's pocket and found itself wedged in the guard's arm. The second rifle tumbled out of the guard's reach and into Ferrière's grasp. As the elevator light flickered back to life, the two stunned guards—one bleeding from the nose, the other holding his left arm and wincing—found themselves facing down a particularly peppery scientist. For what felt like an eternity, all three of them stared at each other in tense silence, only broken by the elevator's jarring ping and the opening of the doors.

I imagine that's close to what played out in Ferrière's mind. The elevator did suddenly stop between the first and second floors, and its occupants were briefly plunged into darkness. "In my brain, I thought: you take a pencil from somewhere, and . . ." he told me, before dramatically mimicking a stabbing motion. But he reasoned that this wouldn't end well for him. He wasn't Jason Bourne. He was a geologist and meteorite curator at the Natural History Museum in Vienna, where he had worked since 2011.[1] He's also one of the world's best impact crater hunters. And on that occasion, his quest got him into a bit more trouble than he bargained for.

I first met Ferrière in person on a chilly January morning. After walking into his office—less of a room, more of a maze of polished wood, books, and geological samples sitting on drawers filled with historical artifacts, including some related to the Austrian emperor, apparently—one of the first things he did was judiciously hand me Austria's piece of the moon. This glassy-looking fragment, one handpicked by an astronaut who sojourned on the lunar surface, was of immense value; it was encased in a bubble of glass, attached to a frame emblazoned with Nixon's signature. Many nations have an official Apollo sample gifted to them by the victors of the first space race. I had previously seen the UK's sample, encased within a cabinet in London's Natural History Museum and displayed as part of a special exhibition. Holding a Moon fragment in my hand was surreal, but, as a somewhat clumsy person, terrifying. I quickly handed it back.

The private tour of the museum Ferrière granted me later that day was full of instantly memorable moments, from walking down a passageway packed to the rafters with skulls, to getting the lowdown on the multiple meteorite-crowded halls, the delivery of which had sometimes been sponsored by famous actors and actresses. But nothing could triumph over the lengthy conversation we had in his office, one filled with tales of being in the field, looking for impact craters that few others dared to try and find.

A monitor in Ferrière's office displayed a stream of seemingly random blips of light. Sometimes, just a few showed up at a time, then a pause, then another. And another. This was a real-time infrasound feed of very small meteors exploding in the atmosphere—a wallpaper of shooting stars. This was a sign of someone who not only loved what they did but wanted everyone else to know about it.

Ferrière grew up in the French countryside, in a small town called Feings. "I had plenty of nature around me and not so many friends living close by," he explained. Like me, he grew up fascinated by volcanoes, but shooting stars also beguiled him, and as early as elementary school he knew he wanted to be a natural scientist of some sort, perhaps a mineralogist, someone who professionally interrogates rocks for a living.

He was taken to church a lot as a child. "I turned away from the church. I have my own church now," he said, gesturing to his museum surroundings. He became fascinated by the adventurous careers of explorers including France's Théodore Monod (who was "the last naturalist in the world," according to Ferrière) and Katia and Maurice Krafft, the Shoemakers of the volcanological community, who dared to go where other scientists wouldn't—until 1991, when an unexpected volcanic explosion in Japan killed them both.

After nabbing several degrees in the Earth and planetary sciences, and jumping back and forth between Europe and Canada, Ferrière eventually landed on impact craters. Impacts are the architects of the solar system. Worlds don't get made without them. Planets also don't get destroyed without them. They are the quintessential love-hate characters of the cosmos, and buried within them are scientific riches. If academics want to get a look inside another world, poking around the solar system's craters is one of the best ways to do it. Want to know the history of our galactic neighborhood? If you can work out when they were excavated, the plentiful craters on Mercury, the Moon, Mars, and all those gigantic asteroids in the belt beyond sketch out the violent symphony that has played over the last 4.6 billion years. And planetary defense researchers love old craters, because simply by counting and sorting them by size, you can approximate how often planets get hit by asteroids and comets.

Unlike the other rocky islands in the inner solar system, however, our own planet is extremely good at cleaning up after a mess. Between its lava-spewing volcanoes, its abundance of life, its flowing water, its abrasive atmosphere, and, most problematic, its jigsaw of tectonic plates—which not only mangles up the surface of the world but also swallows and recycles much of it—Earth lacks the craters of its siblings. It proudly wears a few, including some on unrecyclable continental crust that has been preserved at the surface for billions of years. But many craters older than 200 million years have been destroyed, and those younger than that have been obfuscated by cities, agriculture, ice, trees, and oceans.

It's important that we find them. If we don't, then we cannot confidently say how rarely, or frequently, city killer–, country crusher–, and even planet pulverizer–size asteroids and comets hit the planet. But finding them is easier said than done.

After the Moon fragment, Ferrière placed another rock in my hand. This one, raw and exposed to the elements, looked a little gnarly and chewed up. "It's not every day you have a piece of Chicxulub in your hand," he said. He wasn't wrong: this was like holding a piece of the grave that marked the end of the reign of the dinosaurs. No, it was more like grasping a piece of the explosion itself, a shard of cooled inferno trapped in a collection of grains and crystals—a small piece of such tremendous energy that, if we were able to harness it, it could power civilization as we know it for more than two centuries.[2]

It seems odd that such an era-defining, world-shaking asteroid impact, one that produced a crater 110 miles across and 12 miles deep, proved to be so difficult to find. But most of it was submerged in the Gulf of Mexico, and even when a petrochemical company geophysicist stumbled across it at the end of the 1970s, it took until the early 1990s before scientists confirmed it was the smoking gun—or, well, smoking hole in the ground—that they were searching for. And this was a planet killer asteroid. The smaller ones, those we are chiefly concerned about, are orders of magnitude harder to locate.

Ferrière thinks of himself as a terrestrial astronaut. Like those space-based wayfarers, he can discover plenty of completely new things on Earth, even in the twenty-first century, especially after the satellite revolution of the past few decades. Those increasingly sophisticated eyes in the sky can spy incredibly small things, from pirates trying to steal cargo to the illegal deforestation attempts of corrupt regimes and companies. Satellites can also spot big round depressions in the landscape, some of which can be craters. But there are plenty of circular blemishes on Earth's surface, from volcanic cauldrons to human-made pits.

"You need diagnostic evidence," explained Ferrière—mineralogical evidence. That could include shocked quartz, a type of translucent crystal filled with many small fractures, or shatter cones, groove-covered pieces of the planet mangled by an almighty force. Both tend to come about as the result of two types of explosions: nuclear blasts or meteor impacts. And the only way to get that evidence is to head to any suspected impact craters on foot.

Plenty of the unsubmerged planet has been scoured by scientists hoping to find such diagnostic evidence, and sometimes they succeed. Even the part of the planet that's underwater gets explored: in 2016, a British- and American-led team spent seven weeks above the sunken heart of Chicxulub, drilling into it and extricating its ancient rocks,[3] aiming to learn more about the impact crater and the asteroid that made it. That piece of the crater Ferrière so casually handed to me came from two-thirds of a mile below the seafloor.

Certain parts of the world, however, have been underexplored. These tend to be countries marred by violence and war, caught between colonial pasts and authoritarian presents. It's unwise to travel to these nations. But it seems Ferrière simply cannot help himself. He goes where other researchers won't, hoping he won't share a similarly grim fate to the Kraffts.

The first story he told me was about Luizi,[4] an impact structure on the Kundelungu Plateau, within the Democratic Republic of the Congo. This nation is a volcano-adorned vivarium of biodiversity and natural beauty the size of western Europe. But conflict, corruption, and a cluster of humanitarian crises make the lives of many of its 60 million poverty-stricken citizens a misery.[5] Most international travel advice references the varying levels of instability across the country, noting that foreign nationals have been kidnapped, and that the risk of terrorist attacks is high. And although some highly vigilant travel is possible, it's advised that nobody venture into an enormous swath of the DRC, particularly to the north and east.[6]

Ferrière was going anyway, dangers be damned. In 2010, satellites had identified a ten-mile structure in the southeast that resembled a crater, and he had to confirm its identity. Teaming up with François Lubala and Pierre Kaseti, geologists from the University of Lubumbashi in the DRC, they set out on their mission. The journey was arduous and meandering: roads into the wilderness that showed up on Google Earth did not exist, so they made their own way with 4x4s until even those vehicles couldn't handle the terrain. About twelve miles from the putative crater's edge, they left the car and continued on foot. (Their driver then went on a ten-day joyride before selling the gasoline.)

When the geologists arrived at the crater, they made camp, had a rummage around, and found those diagnostic shatter cones and shocked quartz grains. Together, they had discovered a complex meteorite impact crater—the only one of its kind in Central Africa, according to their 2011 study.[7] This wasn't easy to achieve—but compared to another expedition in 2013, it was a walk in the park.

Satellites had found another round thing in the DRC, dubbed the

Omeonga structure, smack bang in the middle of the country, in the Sankuru province. This possible crater, twenty-three miles long, existed in an area lacking almost any infrastructure, and came up distinctly fuzzy on maps. Kaseti, who had teamed up with Ferrière on his Luizi mission, joined this pursuit, too, though anyone would have forgiven him for not taking part. Ferrière told me Kaseti had previously been captured by armed insurrectionists who detained him for several years, hoping to exploit his geologist training to help them fund their violence with mineral wealth. "His parents died not knowing he was alive." Despite that trauma, Kaseti was eager to help Ferrière reach his geographical target.

Ferrière first arrived in Kigali, the capital of neighboring Rwanda, before traveling to the DRC's sprawling city of Goma. The metropolis is in constant danger of being buried by speedy lava flows from the enormous Nyiragongo volcano[8] looming over it, or being suffocated by an eruption of noxious gases from the shores of the vast Lake Kivu.[9] Outbreaks of armed violence aren't uncommon. But none of this especially concerned Ferrière. What did, at least a little, were the planes. DRC air carriers are banned from operating in several countries because their safety record is so poor that it might as well not exist.[10] He was scheduled to fly from Goma to Kindu, a city to the west, and at the airport in Goma, on the runway, he was greeted by a bizarre sight.

"On the one side you have the lava flow from the eruption," he said. "On the other side, you have a cemetery of airplanes," the remains of wrecked and ruined aircraft, some of which met a fiery end. "You don't really expect that at an airport." He had two flights to choose from. The plane he picked took him without incident to Kindu. The other one crashed not long after it took off. That is deeply unnerving, I said. He shrugged. Look at it this way, he told me: "Half the passengers survived. You must always be optimistic."

In Kindu, his small squad gathered their tools and zipped into the unknown on motorbikes. Finally, they reached the fringes of the rainforest-covered structure. But after they took a long, hard look at their geologic environs, sifting through the grit for days on end, they ultimately came up empty-handed. "I didn't find anything convincing," he told me. No shocked quartz, no shatter cones. It sure looked like a crater. But they couldn't confirm it. You win some, you lose some, he said.

"And now," Ferrière said, "the jail one." He had told me of this misadventure in brief on an earlier call, but I had to know more.

The phone interrupted our chat, as it had been doing all day. He had recently appeared on Austrian TV talking about asteroids, meteorites, and the like—and suddenly, everyone was finding meteorites all over the place that they wanted to report. This, indeed, is the curse of many planetary scientists all over the world: if they put out a call for meteorites after a fireball is spotted, enthusiastic members of the public call in to share their findings with the local meteorite expert. Almost always, it's not a meteorite. But people want to do their best to help—a welcome sign that interest in space is high, though such enthusiasm can transform into an unwelcome distraction for someone like Ferrière. A handful of obsessive callers, convinced that what they have found came from the stars despite no evidence to support that claim, take up more time than Ferrière can handle. "They phone once, I'm nice," he said. "They phone again, I use stronger words." And, I ask him, the third time? He offered an expletive in response.

Anyway, where were we? Ah yes, the time he was imprisoned.

In February 2020, Ferrière and his graduate student, Jean-Guillaume Feignon, were off to Gabon. They were on the hunt for the so-called Bateke Plateau structure, about four miles long, potentially created by a ruinous impact sometime in the last couple of million years. According to Ferrière's maps, the structure was located somewhere close to the eastern border with the Republic of the Congo, itself the western neighbor of the DRC. Some spurts of violence notwithstanding, Gabon is a considerably safer destination[11] than the DRC, and they did not expect to run into any trouble on their field trip.

They flew into the city of Franceville, Gabon, then caught a train, then hopped into a 4x4, then switched to motorbikes. It was a long, long journey, Ferrière recalled—but a relatively uneventful one. Eventually, they arrived at Yabambeti village, right on the rim of the supposed impact structure. He showed me a photograph of himself sitting down outside next to curious but skeptical residents, who were not sure why anyone would travel so far just to look at some rocks. To his surprise, though, someone there told him that he was not in Gabon, but in the Republic of the Congo.

This was a concerning bit of information. The maps of the border region between the two countries suggested they were still in Gabon, but the location of the boundary isn't clearly marked on the ground and appeared to be disputed. The situation felt tense. Ferrière excused himself, saying he was going to clean himself up by the nearby river. While there, he surreptitiously pocketed some sand and silt, suspecting that some of the material may have eroded from the crater-like structure upstream.

Then the police showed up—military police and what also appeared to be the secret service. And they weren't Gabonese. It wasn't clear how they knew Ferrière and Feignon were there, but someone must have alerted the authorities when they saw two westerners sleeping next to a river that, in retrospect, was probably the Gabon-Congo border. That they had not solicited permission from the Congolese government to be in the country meant that their border-crossing shenanigans came across as peculiar. Trying to explain that they were looking for an impact crater may not have helped, considering that they weren't even sure where it was. And Ferrière constantly asking questions at the Yabambeti village also didn't help. "It's probably why they suspected me of being a spy," he told me.

The two scientists were ushered into a military vehicle that then sped them to the Congolese city of Ewo. They stepped outside and were greeted with the grim sight of a military base, whereupon they were promptly chucked into a cell. His student slept on the bed, while Ferrière tried to turn a map into a makeshift mattress for the floor. "I didn't think to escape in this moment. I mean, what're you gonna do?" he said. "I told my student, okay, we are going to be here a long time." Taking risks with his own safety was one thing, but Ferrière regretted getting Feignon mixed up in this unprecedented, deeply distressing situation.

Nobody gave them an official reason why they had been detained. During an early interview, when they were apparently given a chance to explain their expedition, Ferrière witnessed a senior official telling the guard who was keeping notes to instead write what the official told him to write: a fabricated justification for their detention.

Luckily, although most of their belongings had been confiscated, the military police had failed to search the two individuals thoroughly. Feignon still had his phone on him, and it still had both battery life and an intermittent signal. In this situation, many of us would have reached out to our loved ones, but Ferrière advised his student against that. They wouldn't be able to help them out of their predicament, and it would only cause them to panic. Instead, when nobody was looking, he quickly used Feignon's phone to send a WhatsApp message to a friend—the wife of one of France's previous ambassadors to Austria. HELP. IN GABON. ACCUSED OF ILLEGALLY BEING IN CONGO. CAPTURED BY AUTHORITIES, HEADING TO JAIL. ANY HELP WOULD BE APPRECIATED. Send. Fingers crossed.

The next morning, the pair were thrown into another vehicle that took them for a 350-mile cross-country journey of sweat and silence. At no point were they told where they were going. Feignon's phone, still in his possession and in working condition, remained quiet—until, during the drive, it vibrated. They took the call. It was the French ambassador to the Republic of the Congo. "He wanted to speak to whomever was in charge," Ferrière said. The senior officer in the vehicle gave them an icy glare, and said "he doesn't care about the ambassador and doesn't want to speak with him." He forced them to terminate the call.

Several hours later, they arrived at Brazzaville, the capital city. Their remaining belongings were taken from them, they were put in a cell in a grimy, mosquito-ravaged jail, and were kept there for four days. "One of the first things we did is we broke the lights," said Ferrière. "I'm afraid of only one thing there, and that's the mosquitos." The last thing he wanted, when or if he was liberated, was to contract malaria, chikungunya, yellow fever, or any other potentially debilitating infection.

They were not just left to simmer in their own juices. On the second day, they were separated and individually cross-examined by an unpleasant group of interviewers who harangued them for the *real* reason they were there. Ferrière, losing his patience, told them to look him up on Google, where his identity as an impact crater scientist would be easy to find. "And the guy said they don't have any internet, and I was like, shit."

The military implied that Ferrière and his charge were sloppy, supercilious spies. Officially, they were accused of illegally crossing the border. They responded time and time again by explaining that they were conducting geologic fieldwork. But they didn't mention meteorites—a potentially valuable commodity—just in case any of their captors thought that the imprisoned crater-hunting scientists could be exploited for money. By the third day of their confinement in Brazzaville, they had settled in as best they could. For all they knew, it could be weeks, or months, before they stood a chance at being set free.

It was on the fourth day that Ferrière was taken out of his cell, moved outside, and brought to the Ministry of Foreign Affairs. He did fantasize about escaping his armed escort during the erratic elevator ride up to the office of the secretary general. Instead, without incident, he was shown into the office—and was surprised to find not only Congolese officials, but three people who revealed themselves to be from the French embassy. Diplomats, you could call them. Perhaps the French secret service, but no diplomat would cop to that. "The French put a lot of pressure on them to release us," Ferrière told me. The Congolese officials said that nobody could force them to do anything. Like magic, though, that strange gathering culminated in Ferrière and Feignon's freedom.

Before they left the country, someone managed to get their river rock samples back to them, along with two additional rocks from somewhere close to the possible crater. Ferrière doesn't quite know how these samples reached them, but they did—and these rocks are sitting in Ferrière's office, waiting to be examined. Unbelievably, this harrowing tale of incarceration didn't put a dent in his or Feignon's zeal: they had hoped to go back to the area—this time, with visas for the Republic of the Congo, just in case—to locate the suspected crater. But the coronavirus pandemic put an end to those plans. Ferrière told me he still wants to go back sometime in the not-too-distant future, but alone. The only person he ever wants to willingly put in danger, in the name of planetary science, is himself.

Ferrière is in it for the science, sure, but it's the adventure that really lights his fire. He told me about his interest in finding out whether a huge depression in Iraq is an impact crater—a bit difficult, not least because information he had gathered suggested it is used as a dumping ground for "chemical or potentially nuclear shit." If that doesn't work out, there's always another impact crater candidate in war-torn Yemen. His hastily sketched-out initial plan—"Drop me with a helicopter. I'll be thirty minutes, then back up"—was rejected by local fixers, because he'd just get shot down. "I keep it on the list," he said. It's possible that he could put out a call to social media, asking anyone nearby to scoop up some crater rocks and post them to Vienna. "But then you kill all the adventure," he told me. "I like the full adventure."

The death-defying, reckless crater-hunting part of his career aside, Ferrière did acknowledge that contributing to planetary science, and ultimately to planetary defense, was so fascinating on its own that it trumped most other jobs. Like many of Ferrière's scientific colleagues—however they choose to go about it—he's in it for the thrill of discovery.

He handed me a third rock; this one was silvery gray, with a pallid shimmer. "You know what you have in your hand?" he asked. It was a meteorite—but one of the rarest you can find. "It's 165 grams of the Moon."

He was absolutely right. Wonder wins, hands down, every time.

The city of Flagstaff, Arizona, the stepping stone to the Grand Canyon, was in the rearview mirror. Dust whirled around the car as Teddy Kareta zoomed forth, occasionally questioning his own sense of direction. At school, "I was really under this impression that the solar system was this solved thing," he told me. But at the age of nine or ten, "right around when Pluto got demoted," it became clear: "we're still finding new stuff in the outer solar system. We're still figuring out not only where things are, but what any of this stuff is. No one knows what's there, we just know it's weird."

That weirdness seemed like a fun thing to be paid to unveil. Kareta was baffled by the adults who told him that he should enjoy life while he could before he became an adult himself and had to work. Just a few months before our rendezvous in Flagstaff, Kareta had received his PhD. While most adults "have sequestered nine hours of their day to doing something they genuinely hate," Kareta was now a resident small body expert—asteroids and comets, and their strange hybrids—at the city's storied Lowell Observatory, the place where, back in 1930, astronomer Clyde Tombaugh discovered Pluto.

We were on our way to see an impact crater. There are several near-perfectly preserved craters across the planet that anyone can waltz up to and peek at. I had settled on this particular pit in Arizona for a couple reasons: not only was it the size a city killer meteor would make, it was arguably the place where planetary defense, as a concept, was born. Up until that point, I had immersed myself in research on impacts and craters, on what would happen if, say, Seattle was hit by an asteroid with city killing or country crushing potential. I could really visualize it, I thought. Even so, I knew I had to make the pilgrimage to see one myself.

The crater we were about to visit was made by an asteroid, which was a little sacrilegious to Kareta, whose first love would always be comets. Did he have a favorite comet? He paused. "That's something I have strong opinions about," he said, eventually. Most comets are fabulous, but he loves the ones that break from the norm. He declared that 29P/Schwassmann-Wachmann 1 is a top-tier comet. It has a circular orbit, not elliptical like most comets, just exterior to Jupiter's own. And there is no gradual rise or fall in brightness as it nears the Sun and its ices begin vaporizing into a coma. "It's either fully on, or fully off. It has a gigantic outburst, and then nothing." Nobody knows why. But it's weird, and Kareta was all for that.

Pedro Bernardinelli, an astronomer at the University of Washington, is a more recent convert to the cometary church. At the start of his PhD, he worked on the Dark Energy Survey, which charts galaxies and stars to better understand the mysterious force causing the very fabric of the cosmos to stretch at accelerating speeds. Bernardinelli used the survey's deep space photography to look for all the objects photobombing those distant galaxies, including comets in the outer solar system. And in 2021, he became the co-discoverer[12] of the largest comet known to science: C/2014 UN271. The object, now known as Comet Bernardinelli-Bernstein and affectionally referred to as BB, has an icy nucleus eighty-five miles long[13]—almost sixteen times taller than Mount Everest.

Over the past year, Bernardinelli has been lauded by the scientific community. Better yet was the reception he received back home in his native Brazil. "It's not that common for Brazilian scientists to do something with this kind of impact. I feel like I did my part as a scientist, showing Brazilians that they can do science, which is something we don't think about that much in the country. I got to show my country that yeah, we can do science and we're as good as anyone else."

Yet despite this and other discoveries, the first problem comets pose for planetary defenders is that they are incredibly hard to find. They light up when they approach the Sun, yes—but by that point, they are often close to Jupiter and closing in on the inner solar system incredibly fast, perhaps three times the speed of your average asteroid. If one was going to hit Earth, we'd want to see it while it was still beyond Neptune, in the Kuiper belt, or the far-flung Oort cloud surrounding the solar system. But that's a tall order.

Comets are made of various ices, which can be reflective. But they are "dirty balls of ice," Bernardinelli told me. They look a bit like a soiled cappuccino, a soot-smothered soccer ball. They are shady by nature. With the telescopic technology currently operating around the world, many comets are invisible for most of their circumnavigation. To spot them, astronomers rely on them bursting into life, which happens frustratingly close to the Sun-soaked part of the solar system. Scientists may get lucky and spy something with especially volatile ices flaring close to or in the Kuiper belt, as was the case with Bernardinelli's elephantine comet. But it will take an act of sorcery to spy a comet beyond that torus. "The Oort cloud sometimes freaks me out," confessed Bernardinelli. "It's so far from the Sun, and we only know about it because some things get kicked in and we see them."

"I'm just going to tell you the truth," Kareta told me, as we approached our turnoff in Arizona. "Comets, in the grand scheme of things, are more likely to cause an extinction-level event." They are stealthy, often dozens of miles long, and move at terrifically high speeds. Comets can also exist on highly inclined orbits, putting them in parts of the night sky that are difficult to observe. Sometimes they orbit the Sun in the opposite direction to Earth, so if we get astoundingly unlucky, we "could get a head-on collision at seventy kilometers a second." (That speed would get you from New York City to Miami in twenty-five seconds.) "Are you starting to see the picture here?" Kareta exclaimed. "When it comes to doing something like DART, we are protecting against the kind of apocalypse we can envision."

Could you, I don't know, explode an incredibly powerful nuclear weapon next to a comet and deflect it? According to Megan Bruck Syal, our NED expert on the planetary defense team, countless experiments she and her colleagues have conducted at Lawrence Livermore show that that option doesn't look great. "Comets are, by far, the most challenging scenario. That's kind of the nightmare scenario, the most difficult to mitigate," she told me. Helpfully, X-rays from nuclear blasts penetrate ice more deeply than rock, so theoretically you can get a bigger push on a comet compared to an asteroid of a similar size. But "we're unlikely to see a long-period comet more than eighteen months ahead of time."

That's the bad news: if a comet were heading toward us, we are probably doomed to die. The good news is that, because comets are in the distant outer solar system, and only rarely pop into to our inner sanctum to say hello, the chances of being hit by a comet in our lifetime, or in the next dozen lifetimes, are infinitesimally small.

Asteroids, more commonplace and hanging out much closer to home, are orders of magnitude more likely to hit Earth. But at least we can do something about them. That's what DART is hoping to do. Kareta, for his part, always felt confident about our chances with asteroid deflection. "We evolved from apes," he said. "We know how to throw shit."

He squinted at a road sign as we approached. "I think it's the next exit," he said, adding unnecessarily, but amusingly, "I'm not kidnapping you." He was right on both fronts. We had arrived, and soon the ground would fall from beneath our feet under the hot Arizonan sun.

If you look through a telescope from the raised rim of Meteor Crater, you can see a life-size astronaut cardboard cutout placed at its center. Even with the telescope, it's difficult to see the cutout: the crater, carved out of the Arizonan desert 50,000 years ago, is almost a mile wide and about 600 feet deep. London's St. Paul's Cathedral or the Washington Monument would fit comfortably inside, with their summits not quite managing to reach the same height as anyone standing on the rim.

Standing before a small crowd, including Kareta and myself, the guide explained that "this event was similar to a nuclear blast going off." The asteroid was made not of stone, but iron, making it an atmosphere-piercing bullet—300,000 tons moving at 26,000 miles per hour. When it hit the ground, at least 80 percent of the asteroid was vaporized. I had glanced at a surviving fragment inside the welcome center on my way to the crater. It had the appearance of ink-black liquid metal kept in suspended animation, and although it was just 2.5 feet long, it weighed as much as an adult polar bear. It looked like something the followers of Sauron might use to wield dark magic.

The asteroid it had come from was 140 feet across, perhaps a little smaller than the Tunguska impactor. But it would have taken only three or four seconds from the fireball first appearing in the sky to it plunging almost directly downward and turning the earth into gas and plasma. In a heartbeat, the local greenery would have suddenly become nothing more than light and rage. Over thousands of years, the crater has been infilled with sediment. But originally, the wound was twice as deep as it is today: 1,400 feet, enough to conceal the Eiffel Tower with headroom to spare. "It had a kill zone radius surrounding the crater of seven to eight miles. Disintegrated. Destroyed. Nothing survived within that zone," the guide told us. Large chunks of debris, the size of houses or bigger, were thrown up to forty miles in every direction, about as far away from this spot in the middle of nowhere to the city of Flagstaff. "If you look behind you, it barely missed our visitor's center," the guide said, with a hearty wink. The crowd chuckled. I was too distracted; I couldn't stop thinking about that horrifying, eight-mile death radius.

Back then, the landscape here was not arid, but lush: open grasslands and assorted woodlands that were home to all sorts of animal life, from mammoths to giant ground sloths. "At least it would have been a quick death for most of them," Kareta said.

The guide then gestured toward the crater rim across from us. There I saw something that I had seen plenty of times during lectures back when I was a university student but had never seen in person: the very ground we walked on, and several geologic layers beneath it, were warped and turned upside down, flipped over by the impact like waves of water moving away from a dropped boulder. "This is the place where the impact and the explosion were strong enough to not break the rocks, but bend them," Kareta said, giving those last words weight. When solid rock flows like a liquid? "That's bad!"

It's no wonder that this is where the first scientific seed of planetary defense was sown. The guide explained that the crater was first found by Native Americans, and was recorded afterward in writing in 1871 by the US Army. "A scout was declared delusional when reporting this area," he said. The name of the scout was mockingly used to dub the crater Franklin's Hole. After the existence of this enormous pit was confirmed by multiple parties, it was thought for some time to be a volcanic crater, a place where pressurized molten rock or a giant pocket of magma-heated steam had triggered a huge explosion.

No volcanic evidence was ever found, but the notion persisted, because the alternative—that something terrifying fell from the heavens—seemed too absurd to be true. The telltale iron chunks around the crater may have steered people in the right direction, but they were not handed to scientists. "The railroad company bought those pieces and forged the railroad," the guide said—the foundations of locomotive transport, built on alien metal.

In the 1920s, Daniel M. Barringer, a mining engineer and businessman, thought that a profit could be made from the crater itself. If iron shards were around this crater, then perhaps a vast store of the stuff was buried beneath the pit's heart. Today, the fifth generation of the Barringer family still own much of the area's land—but they never did find that iron, and mining operations were abandoned in 1929.

To his credit, Barringer did buy into the impact theory, and he was unopposed to sharing the site with open-minded scientists. This included Harvey Harlow Nininger, one the first academics to professionally study meteorites, who documented peculiar geologic features, including rock that was both shattered and fused together by some huge force.[14] In a report from 1964,[15] Barringer's son recounted the changing tides: "By the time of father's death, November 30, 1929, the great weight of scientific opinion had swung around to the accuracy of the impact hypothesis."

US Geological Survey maverick Gene Shoemaker, future co-discover of the comet that would turn planetary defense into a realistic pursuit, came along in the late 1950s and started sleuthing the crater. Shocked quartz, the very same matter Ferrière and his contemporaries rely on to confirm possible impact craters, was identified in pieces of ejected sandstone.[16] And in July 1960, Shoemaker and his colleagues Edward Chao and Beth Madsen announced that they had found a mineral called coesite,[17] odd crystals that can also be forged during nuclear explosions. That was definitive proof that the chasm was made by something similarly energetic: an asteroid impact. In 1963, it was made official: this, the guide proudly declared, was the first proven meteor impact zone on Earth. And the harsh reality that Earth remains vulnerable to meteoric misfortune started to creep into the scientific community's consciousness.

Asteroids are aesthetically majestic, in their own explosive way. But that day in Arizona, it became gut-wrenchingly clear to me that destruction is their modus operandi. I intellectually knew the threat that an asteroid impact posed to the planet, but that afternoon, I felt it—that anxiety, a palpable understanding that if we don't confront this cosmic challenge, millions of people will one day be condemned to death.

"DART is like a rock-hammer-in-space-style experiment. Let's go and smash something and see what it looks like," said astronomer Nick Moskovitz, leaning back into his chair. "That's almost cave man–level innovation. And that's where you start with planetary defense. You've got to start somewhere."

By July 2022, DART was closing in on its target. September 26 was the lock-in date for the experiment's big swing, one that would either work beautifully or wouldn't work at all. Moskovitz, speaking with me after my visit to Meteor Crater, was a member of Cristina Thomas's international team of observing scientists. At the Lowell Observatory atop that leafy Flagstaff miniature mountain, he explained how Didymos and its mostly invisible moonlet, Dimorphos, had been in their sights since 2015. "You need to know where Dimorphos is to hit it with a spacecraft. And it's our observations with the Lowell Discovery Telescope that have contributed to that."

Thomas's seekers, including those at the Lowell Observatory, had just completed a six-night observation campaign,[18] one that affirmed earlier campaigns' efforts to track the binary asteroid system across the night sky. Like many of his colleagues, Moskovitz was positively thrilled to be part of a mission that was not only noble and deeply cool—but one on an accelerated timeline compared to the usual, glacial pace of such missions. DART had "tendrils going back" decades, but from "green light to completing the mission is, like, seven years. That's crazy for a planetary mission. I really like that. I'm by nature somewhat impatient; I like how fast this is moving, and how streamlined it is."

I suspected Thomas herself, although also enjoying the breakneck evolution of the mission, might have wanted to freeze time on a few occasions. She had spent much of the warmer months making sure her global-scale organizational powers were on point. She had the spreadsheet to end all spreadsheets, a roster of the planet's observatories, explaining who would be doing what to watch Dimorphos get its bell rung at that crucial, blink-and-you-might-miss-it moment. "I spent a lot of the summer getting all of this ready," she told me. "It was completely absurd."

After surviving that mild case of electrocution involving the hyperactive NEXT-C thruster, Elena Adams and her team of systems engineers had been going through the spacecraft's travel photography. On July 1, and again on August 2, DRACO was pointed not at Didymos, but at Jupiter and one of its moons, Europa,[19] an ocean-flooded world completely covered in an icy carapace. This wasn't (just) a bit of interplanetary sightseeing. The team was putting the robot's brain, SMART Nav, through its paces. As the moon emerged from behind its parent planet during its orbit, the software was given a test: it was asked to switch its focus, all by itself, from Jupiter to Europa. If it could, it would bode well for the critical moment in the last hour of its life, when it would be required to find Dimorphos sneaking out from Didymos' shadow.

"We fooled our SMART Nav," Adams told me. The test worked wonders.

But something else was fooling them. DART's star tracker, the instrument that told the spacecraft its exact position in space, had been acting strangely since April. It kept telling the ground team that DART was somewhere it wasn't—close, but not close enough. They were troubleshooting the problem but hadn't yet come up with a fix. And the longer this technological illness was left to metastasize, the worse the spacecraft's prognosis would be.

Publicly, the news updates about DART were all positive. But behind the scenes, Adams and her team knew that there was a very real chance that their spacecraft might lose its way out in the star-filled sea. How likely was it that DART would hit Dimorphos? "We were pretty honest with NASA during the summer that we had about a fifty-fifty chance of hitting," she told me. Just two months before impact day, the creation of Earth's planetary defense program rested on the flip of a coin.

VII

A Dress Rehearsal for Saving the World

On March 11, 1998, Dennis Overbye—the new deputy science editor at the *New York Times*—walked into his afternoon news meeting. "We've got a new story," he told those around him.[1] "It's a pretty good story. It's about the end of the world."

In the summer of 2022, I sat next to him at a café at the southern edge of Harlem. Now a science journalist for the same publication—he realized he preferred telling tales as a bard, not organizing others as a conductor—Overbye has penned stories about as many aspects of the universe as you can imagine, from collapsing neutron stars to pinwheel galaxies emerging from the haze after the big bang made its mess. But he keeps that day back in 1998 in a special place in his memory cabinet. After all, how many times do you get to walk into an editorial meeting and tell everyone there's an asteroid coming to kill them?

Was he worried at the time? "We're all going to die," he told me. "Get over it." A good story's a good story, right?

One fine day in 1997, a mile-wide asteroid was found by astronomers. They watched it for a while and connected the orbital dots. Uh-oh. The rock, easily capable of squashing a country or two, looked like it was wandering our way. They didn't have many observations, so there was a fair bit of uncertainty. But Brian Marsden, director of the Minor Planet Center, the bulletin board for all new NEO discoveries, thought the trajectory was alarming enough to warrant letting others know[2]—including the press. Information about the asteroid, dubbed 1997 XF11, and its projected pathway through the solar system was sent out via postcards and telegrams in March 1998, with one reaching Overbye's reporter colleague Malcolm Browne.

The notice did not come adorned with bright red exclamation marks, so it was not immediately obvious what it revealed. But Overbye recalled seeing that the distance between the asteroid and Earth "goes to zero at some point." It was not quite zero: on October 26, 2028, the asteroid would pass within 30,000 miles of the planet, which is about 13 percent of the distance to the Moon. But a little uncertainty could mean that the distance would ultimately become zero. "Well, that doesn't look good," he said to himself at the time.

After a quick and thoroughly unusual conversation with Browne, Overbye stepped into his 4:30 p.m. meeting. He had been in the job for only a month, and his boss was away that day, leaving him in charge. So, what might be on the front page of the *New York Times* tomorrow? The weather, in the top-right corner, of course. President Clinton moving his China visit to late June, sure. Oh: that little thing about the world's close shave—or lethal rendezvous—with a huge asteroid.

In 1998, both astronomers and the media were wading through murky territory. Who should be told? How should they be told? The meaty communication systems in place today were skeletal back then, so everyone was improvising. CNEOS had not been inaugurated. The notion that close flybys happen all the time had not yet crystallized in the minds of the general public. The rollercoaster rises and falls in impact odds for newly discovered asteroids was a novel concept for many. There were no clear antecedents to lean on. On March 11, during that editorial gathering, the possibility of 1997 XF11 impacting Earth in 2028 remained just that—a possibility—"but it was in the range of probability," Overbye told me. "So, of course. This had to be a front-page story."

He and his staff spent the entire night in the offices, gathering sources, producing graphics, double and triple-checking their facts. They got a lot of attention from journalists in other departments. Some of them semi-jokingly asked if they should continue to pay their mortgages. (They were advised that they should.) When the dust had settled, the March 12 edition of the *Times* printed with Malcolm Browne's article, headlined "Asteroid Is Expected to Make a Pass Close to Earth in 2028"[3] above the fold on A1: "An asteroid is likely to pass within 30,000 miles of Earth on Oct. 26, 2028, a Thursday," it said, "and there is a possibility that it would hit Earth, the international astronomical agency that tallies the orbits of asteroids and comets announced yesterday." (I love that they mentioned it was a Thursday—at least nobody's weekend would be ruined.)

Marsden, quoted in the story, pointed out that the asteroid could hit Earth, but it could also "come scarcely closer than the Moon." He was half-right. By the time the story was published, astronomers had found the asteroid in older images, giving them plenty of extra observations to refine their calculations. The numbers showed that 1997 XF11 would comfortably miss Earth by 600,000 miles.[4]

Marsden did not come out smelling of roses. He was admonished by several in the astronomy community for throwing out an imprecise impact possibility—one that was in dire need of more observations—into the public realm without consultation. "This caused a big kerfuffle, if you want to use that lovely word," Paul Chodas, the director of CNEOS, told me. Like his colleagues, he admired much of Marsden's work. But he shook his head as he recalled that incident. "He said it would hit, and we said it wouldn't hit." It suggested to the public that astronomers were not as erudite as they seemed—and, in the end, this incident helped spur calls to formalize the NEO observation team at NASA into CNEOS.

Marsden, who died in 2010, may have erroneously—some would say irresponsibly—jumped the gun. But 1997 XF11 was an education for everyone. Overbye's team was reporting on a new sort of science story that merited serious consideration. The impact's uncertainty was included in the article, and everything checked out at the time. The team, and the readers of the *Times,* were thrilled to learn that a future Thursday was safe from an asteroid strike, even if they were annoyed that they had to keep up their mortgage payments.

About 30 years ago, astronomers were unsure of how much NEO information to share with the wider world. Richard Binzel, inventor of the "should we freak out about this asteroid or not" Torino Scale, remembers discussions among scientists where they fretted about when, exactly, to inform the planet's many billions of a possible upcoming impact event. Unless they were 100 percent certain, nobody quite knew when or how to sound the sirens. But stories like 1997 XF11 made it clear: "transparency is the best path to public confidence," he said. The best course of action was to make all the NEO data public, as it is today. "The sky is open to everyone. Do look up and see for yourself. This is knowable, and independently, by any astronomer anywhere in the world." If organizations like NASA translate the data into something digestible and colloquial, if they liaise with the press and the government to make sure nobody frets unnecessarily, then there will be nothing to hide.

Overbye was in his midfifties back then. Now, he mostly thinks back on that strange story fondly—particularly the moment he announced the world may end. "It was fun to say," he said, after some thought. It was also darkly bewitching to think of Earth as not just an observer of the cosmic chaos, but a participant. Today, Overbye is firmly on the wonder beat. "I like stories with big narrative arcs," he told me. And perhaps 1997 XF11's arc is unfinished. That asteroid is harmless. But eventually, should our luck slip away before we are capable of defending ourselves, one of its siblings may not miss Earth by 600,000 miles. Instead, the arc will be completed, and a city or country gets destroyed.

What would that be like to live through?

Everyone was ready: NASA, nuclear weapons scientists at the Lawrence Livermore National Laboratory, astronomers, US Space Command, the Department of State, the Department of Defense, the Federal Emergency Management Agency, even the White House. A killer asteroid was barreling toward North Carolina and there was nothing anyone could do to stop it. Impact was imminent. In just a matter of hours, America, and the world, would be transformed through an act of cosmic violence. Everyone hoped for the best but braced for the worst.

And that's when Russia invaded Ukraine.

Few people knew how to react. But Leviticus "L.A." Lewis did. "Guess what? The asteroid won't care that humans are trying to kill each other. It's still going to keep coming, and we're going to have to respond." Nobody could disagree. Lewis, who is the representative for the Federal Emergency Management Agency, or FEMA, at NASA's Planetary Defense Coordination Office, turned to the hundreds of assembled federal, state, and local officials, scientists, engineers, and emergency managers. The show must go on, he declared. And so, it did: on February 24, 2022, as the Russian military launched their assault on their democratic neighbor, North Carolina was hit by an asteroid.

Between February 23 and 24, while DART was on its way to Dimorphos, experts had gathered at its birthplace, the Johns Hopkins University Applied Physics Laboratory, for a role-playing game. Everyone assumed the same character roles they had in real life. Their realm was not mythical, but terrestrial; their nemesis was not a dragon, but an Earthbound asteroid.

This game had been played several times before—sometimes more casually, sometimes in the United States, other times abroad, each with a varying number of participants. But this iteration of the war game was the most detailed, the most populated, the least imperfect simulacrum of the worst disaster yet to transpire.[5] And Lewis was one of several dungeon masters, watching the players respond to a preplanned but malleable story. The goal: to see how people handle the unthinkable, and to hear how they may ward off the direst outcomes—all to prepare them for the day that this grim fantasy becomes a reality.

The scripts always change with each game. Sometimes, the detection-to-impact timeline is years; other times, it's just a few months. The asteroids can be enormous, or Tunguska-size threats. They can impact land, or bodies of water, anywhere on the planet. Whatever the situation, the timeline is compressed and squeezed into a few days, in which players try to handle the many surprises the dungeon masters periodically lob into their gameplay.

CNEOS director Paul Chodas (along with his team) was chiefly responsible for coming up with the rocky dragon's stats and its Earthbound trajectory. When I spoke with him, he quickly dismissed the dungeon master moniker. "For me, it's a mathematical problem," he said. Indeed so. Then again, to be fair, like a dungeon master, his task is to work out how much damage a Demogorgon (an asteroid) can inflict on a mage (us). "Originally, it was an instructive exercise to inform the community, and decision-makers especially, of what we would know if there was an impact scenario—what the uncertainties would be and how our knowledge would evolve over time and the limitations of our knowledge, frankly."

That aim had not changed over the past few years. But the intensity of the exercises, and the number of people playing, had risen considerably. In February 2022, the games had reached their pedagogic zenith—and much of that was down to the leadership of Lewis, the FEMA representative, someone who never quite believed he was handpicked to do battle with otherworldly forces.

He was a kid when President Kennedy gave his famous 1961 speech[6]—"ask not what your country can do for you, ask what you can do for your country"—and it resonated with young Lewis. Public service seemed appealing. But he also loved things that went really damn fast. "You know, *Thunderbirds, Captain Scarlet,* all of that crap, that was me, I was all over it," he told me, every word animated, emphasized, bursting with zeal. And as he emerged into adolescence, his passion was kept alight, aided by the Apollo space program.

"The most powerful weapon my mom got for me was a library card. It was all over after that," he said. "I was always nerdy, always liked that stuff. Everybody else in my neighborhood went to the basketball court. I got my dad to get me a telescope." He attended science fairs and launched model rockets. He was teased—but his smarts eventually won him the respect of others.

Being an astronaut, or building rockets for spacers, was his goal. Instead, his career became something more kaleidoscopic. He served twenty years in the US Navy, studying the ways of ship-based surface warfare and ultimately reaching the rank of commander.[7] He retired his commission and was working at the Pentagon in 2001 when the September 11 attacks took place. In 2007, he joined FEMA, taking on multiple high-profile roles at the nation's disaster coordination agency while working closely with the FBI. He got married, had kids, and after forty-four years of service, he felt he was approaching retirement. Hanging up his cap on Christmas Day, 2023, felt right.

But a few years before his planned retirement, NASA gave Lewis a call and told him he was needed for one last mission: dealing with asteroid impacts. "I was ecstatic!" he told me, laughing. What better contribution to the world, he thought, than helping to protect everyone on it. "It'd be nice to know I contributed something to this place before I check out."

FEMA normally deals with terrestrial natural disasters—hurricanes, floods, and so forth. Asteroids may be out there in more ways than one, but they are still a type of natural disaster. "It's another low-probability but extremely high-consequence event. It's responsible for us to be prepared for it," said Lewis. And he and his fellow dungeon masters pulled no punches with the 2022 war game: the discovery-to-impact timeline was only six months. It was going to hit America this time, though initially, they weren't quite sure where the strike would occur.

"In this case, a municipality in the United States actually volunteered to be the impact victim," he told me: the North Carolinian city of Winston-Salem,[8] population 250,000. And for the first time, hundreds of state and local officials—not just mostly federal officials—were involved, dialing in to the Applied Physics Laboratory in Laurel, Maryland, from their home state. Isn't it sort of funny that a city offered to be the victim of a simulated catastrophe? "I'm not sure some of their fellow officials thought it was funny," Lewis said.

Plenty of the participating local officials and first responders didn't know what to make of all this. Why would anyone offer up their metropolis as a sacrifice to the astral gods? You would have to ask the guy responsible: August Vernon, the city's director of emergency management. "I'm not a scientist," he told me. "I'm a closeted sci-fi geek." And he was euphoric from the word go. In a manner of speaking, he couldn't wait for Winston-Salem to be pulverized. It's true that, early on, his colleagues were either skeptical or outright hostile toward the war game. "Is this a joke? What are we doing this for? What's next, aliens?" he recalled them asking. It wasn't a totally unreasonable reaction, considering how many other problems—including the coronavirus pandemic—they were actively addressing. But like Lewis, Vernon considered an impact to be like any other tragedy: everyone thinks the government is overreacting until something awful happens, and then they ask why the government didn't do more to prevent it.

Every year in Winston-Salem, city and county emergency management officials get together and run various drills. "From school shootings, to cyberattacks, to plane crashes . . . we had done pandemic exercises before COVID," Vernon said. "We have what's called an all-hazard approach. The federal government does the same thing." Big eruptions, earthquakes, hurricanes, chemical plants blowing up—you name it, they've rehearsed it. An asteroid impact was the logical next step. And why should they take the Hollywood route? "Whether it's an asteroid or a Godzilla attack, it doesn't matter," he said: it's always the same major cities getting destroyed. Why should Los Angeles get all the apocalyptic attention? Vernon reached out to Lewis, and the simulated asteroid had its North Carolinian bull's-eye.

Vernon, a keen observer of past planetary defense role-playing games, wanted to make an important tweak. Previously, the players had been mostly scientists. The real world contains mostly nonscientists. "We didn't want to do an exercise with just what we call ten-pound brains," he said. Winston-Salem could be destroyed, but only on the condition that a diverse range of local officials could be players. Lewis and the dungeon masters eagerly agreed.

"It's the first time in my life I've worked with the Applied Physics Lab," Vernon told me. As excited, though, as he was to work with actual rocket scientists, his local community required a little more convincing. "I had to gently ease them into this," he said of an incomparably dramatic and terrifying disaster. And on February 23, North Carolina and Maryland tried to stop it happening. The games had begun.

Day zero.[9] America didn't know what was coming its way, because when the hypothetical asteroid 2022 TTX was discovered by a NASA-funded survey on February 11, 2022, it looked almost harmless.

In this alternate timeline, astronomers from all over the world peeked at its observational data—a dozen sky positions recorded over two nights—posted up on the Minor Planet Center's bulletin board, and nobody found any reason to be concerned. CNEOS's Scout ruled out any impact in the next thirty days, so it was passed on to Sentry. It determined that there was no risk to Earth with one exception: six months into the future, on August 16, there was a one-in-2,500 chance of an impact, which is no different from many newly identified NEOs. But, just in case, astronomers kept up their observations of the faint speck as it scooted across the dark, a distant 37 million miles from home. After feeding this observation data to Sentry on February 16, the impact odds rose to 5 percent.

A 5 percent impact probability was somewhat uncomfortable. That the collision might just be half a year away was disturbing. And with just a smattering of observations, nobody knew where it might hit. The range of possibilities covered two-thirds of the entire planet, including the Americas, Europe, Africa, and most of Asia. If it did find Earth, it might land in the ocean. But it might not. And based on how much sunlight it was reflecting, 2022 TTX was roughly 330 feet across—a city killer, for sure. But it was also so faint and distant that astronomers couldn't quite tell if it was small and very reflective, or big and not that mirror-like. It was possible it could be 100 feet across, a small Tunguska. It could also be 1,000 feet long, a comfortable country crusher.

It's an unenviable situation to be in. Think of it this way: you are faced with twenty doors, and nineteen of them are safe to open. But one of them, if opened, will free a monster that will devastate a random location on Earth. How confident do you feel that you will open the right door? There was, however, an element of calm at this stage. The asteroid had not yet been observed across enough of its solar orbit for scientists to know its trajectory with precision. "It is not yet possible to predict whether future assessments will indicate the asteroid will miss the Earth or hit, but the chances that the impact will eventually be ruled out are high," the official report on 2022 TTX noted.[10]

At this point during the exercise, a thought popped up: When do you wake up the US president in the middle of the night? This has been shown in movies plenty of times, and it does happen in real life when a huge disaster strikes anywhere on the planet, or when something seismic happens in the geopolitical realm. What would happen with an asteroid impact scenario? What impact odds, and what time-until-impact, would require someone to rush into the second floor bedroom of the White House to prepare the POTUS for a truly rude awakening?

I asked Kelly Fast of the Planetary Defense Coordination Office. It turns out that the president or their senior staff would have likely already found out through that most twenty-first century of sources: social media.

All the Minor Planet Center's information is public. Professional astronomers are not the only ones who browse its notices: countless "amateur" sky-gazers all over the world regularly check it out, and many can run the numbers accurately enough to determine any consequences. "You can't hide the sky," Fast said. "If something should be found that poses an impact threat, chances are it's going to be all over Twitter before any of us gets a chance to say anything." On almost every occasion, denizens of the internet making cataclysmic forecasts are either deluded or morally bankrupt. But there is a version of reality in which the president finds out the planet is in trouble because someone online said so, and that someone just happens to be an astronomer.

Either way, Fast explained, impact threats are taken seriously, and the cogs of planetary defense spin up to a whir when there is a 1 percent impact probability in the not-too-distant future. "That's really the threshold." And should social media not make a mess of things, there is a procedure for who should be told, and when.

NASA Policy Directive 8740.1, as it's catchily known, broadly spells it out.[11] Senior staff at NASA's Planetary Defense Coordination Office (PDCO), having got the word from CNEOS of a worrying impact possibility in the coming months or years, relay that news to the NASA administrator. The call (perhaps just the NASA administrator, possibly with the PDCO on the line) then goes out to the National Security Council, the president's retinue of national security, military, and intelligence advisors, and several cabinet officials.

The news is then filtered down through the many branches of the federal government. The president's national security advisor, buoyed by the National Security Council and in consultation with the director of the White House's Office of Science and Technology Policy, coordinates the American response. FEMA is informed, and military leadership get introduced to a threat very different from those they are used to confronting. Naturally, NASA directs the show, asking for support from the other organs of state when needed. The president's national security advisor acts as a bridge between NASA and the president, who makes the big picture decisions.

With 5 percent odds of a city killer impact in just six months, it's inconceivable to me that, at some point, the president would not make a statement to the press. But in truth, no one is sure how this would play out, because such a scenario has never unfolded in the modern era. In 2013, astronomers tried to preempt all this when they convinced the United Nations to establish the International Asteroid Warning Network, or IAWN.[12] The network has many roles, but its primary goal is arguably to become a globally respected source of information on hazardous asteroids and comets—and to be the place to which people, including governments, will be drawn when a situation like 2022 TTX arises.

So, when does IAWN send up a flare? If a thirty-foot NEO is found, and it has a 1 percent chance of hitting Earth, it will publish a placid notice. But if it's twice that size (a Chelyabinsk-esque object), and the odds of an impact within two decades are 10 percent, it will send out a warning, and advise that "terrestrial preparedness planning" should begin. If the object is Tunguska-size, and there is a 1 percent chance of impact in the next fifty years, it will advise that UN member states with space agencies should set into motion serious space-based defensive options.[13]

These thresholds are somewhat arbitrary. Nobody is quite sure what things like "terrestrial preparedness" would mean in practice. "We're really just trying to get people aware," Tim Spahr, the current manager of IAWN, told me. The network would assist and advise those in North America and Europe. But much of the rest of the world would be somewhat in the dark when it comes to any attempts to mitigate asteroid impacts. In the event of something like 2022 TTX, IAWN would get the other continents up to speed.

In our alternate timeline, by February 23—just a week after the asteroid was discovered—the entire planet was experiencing a novel type of trepidation. With more observations plugged in to Sentry, the odds of impact on August 16 had jumped to 71 percent. We now had ten doors available to open, and seven had monsters behind them.

The asteroid might have closed the distance a bit to now be 34 million miles away, but it was coming on fast. Extra observations, including some images of the sky captured before February 11 that were later found to include the asteroid, reduced some of the uncertainty about its trajectory: much of the planet was safe from direct harm, but the mid–South Pacific Ocean, the mid–South Atlantic Ocean, and the continental United States were all possibly in the firing line.[14] That's when the dungeon masters threw a curveball: misinformation about the asteroid was being disseminated across social media, because of course it was.

What could the government do? "You have to be constantly out in front," Lewis told me. NASA would try to be the trusted face of the crisis. Responsible journalists would convey the correct information in a compelling, clear manner. "But are you gonna convince everybody? No," he said.

Vernon let out a long sigh and took a moment to think. His excitement at taking part in this simulation temporarily withered. "I don't like social media. I think it's terrible, what it's doing to our kids. I know there are positives with it, but in my line of work, all we deal with are negatives."

During the coronavirus pandemic, there was an eruption of misinformation from every possible corner of the media landscape, from cable news and public radio to everyone's online cousins or uncles. Millions were convinced that COVID-19 wasn't worth worrying about, or that masks were useless, or that the miraculously effective vaccines would give them webbed feet. People died when they should have lived. Imagine being an emergency manager trying to put a stopper on that hosepipe of nonsense. "I had no idea there were so many Facebook disease experts in infectious diseases in our community," said Vernon. "When COVID hit, it became overwhelming. Some people, it's the only place they get their news from anymore." The ghosts of the pandemic still haunted him and his colleagues. "I can't tell you how crazy it got." Now, he told me, imagine what that would be like if everyone in the world is told that we are likely to be hit by a dangerous asteroid in a matter of months.

"Forty percent of people don't believe the government at all anymore about anything," said Vernon. He brought up the concept of truth decay,[15] possibly the most disturbing wrinkle of social life that appeared in the 2020s, which is partly defined by an escalating inability for large communities to agree on objective reality. Many things in life can be effectively debated by rational people with informed opinions. But when people are irrational or are acting in bad faith, and their opinions are uninformed or misinformed, all bets are off. The comedian John Oliver wryly illustrated the point in 2014, when talking about the mere existence of climate change: "You don't need people's opinions on a fact. You might as well have a poll asking: 'Which number is bigger, fifteen or five?' or 'Do owls exist?' or 'Are there hats?' "[16]

You may hope that an asteroid plunging toward Earth would be the sort of crisis that's impervious to idiocy. But it won't be. And no effective solution to misinformation has emerged. It's like a superbug resistant to all our antibiotics, moving through the human ecosystem faster and more pervasively than any attempts to stem it with the truth.[17]

No matter how gravely and succinctly NASA, ESA, JAXA, IAWN, and their compatriots share the facts, the nonsense will be overwhelming: the asteroid is a false flag; they just want to take away our freedoms!

Unsurprisingly, at that 2022 interagency war game, nobody quite knew how to handle the spread of falsehoods. But the plague of misinformation is so serious that it made emergency managers daydream of desperate options. One option brought up briefly, and very delicately, would prove to be controversial: shut down social media access by government order. As Americans, that "makes us a little itchy," Vernon said. "We don't like those things. But if the false information is killing people . . . ," he trailed off.

Speaking of desperate measures: the dungeon masters asked everyone how they felt about NEDs—you know, using nuclear weapons to try and stop 2022 TTX, just in case the impact odds got even higher. Most people present didn't really know how to psychologically process that option. Politically, the idea of placing a nuke atop a rocket or a spacecraft and piloting it into an asteroid is fraught with tension. There is no clear international framework in place, no flow chart blessed by the UN that says which nation will be in charge. Obviously, only those nations with space agencies can help in this way. But it just so happens those nations tend to own nuclear warheads. But who gets to launch the mission to try and save everyone else? Does it depend on which nations are at risk? It's not difficult to imagine a scenario in which America announces that they will launch a NED at an incoming asteroid like 2022 TTX, but China chimes in and says that they will give it a go, too. Will they work together, or is one a backup option? It would be mayhem, a nuclear arms race to defend the world.

Nobody wants to use nuclear weapons for any reason if they don't have to. But what else could you do with just six months until impact? "If I was the president, I would expect NASA to give me options, no matter how remote they're going to be. Most citizens are not going to accept that there's nothing we can do, let's just take the hit," Lewis told me. "There would be international implications for this. But when you narrow it down to your country, you start to think about other things, too, like the right to self-defense. I'm not going to wait for the UN or somebody else to vote that, eh, we've given the United States permission to use any means necessary to take care of this. You don't take anything off the table to prevent an asteroid impact that could devastate entire sections of your country or take out an entire city. There's no idea off the table as far as I'm concerned."

Nick Moskovitz, our Didymos and Dimorphos observer at Arizona's Lowell Observatory, told me that the public's perception matters here. NASA isn't militaristic. But the use of a NED, he said, means "you're talking about Air Force, or Space Force... is that what Space Force does? I don't know."* In this situation, NASA and the Department of Defense would be working hand in hand. Many would balk at the thought, especially those outside America. That means it's important that choosing nuclear disruption or deflection is "not viewed as people in camo behind computer screens."

Past tabletop planetary defense exercises have seen participants openly wondering if the Department of Defense should be more integrated with NASA. It may sound like a corruption of NASA's civilian constitution and its philanthropic, scientific ideals. But it is not an entirely unreasonable suggestion. "The DoD is interested in all forms of defense. And this is a planetwide form of defense," said CNEOS chief Chodas. "DoD has launch vehicle capabilities, [they] certainly [have] rapid response launch vehicle capabilities that they're exploring." DART itself was launched not from a NASA facility, but a Space Force base in California.

---

* Broadly speaking, the role of the Space Force is to defend American operations and assets (like launch vehicles and satellites) in space—specifically, the part of space that's very close to Earth. Dealing with potential threats farther out, like asteroids, is not currently one of their objectives.

Such conversations are embryonic. And nothing changes the fact that the US military does not concern itself with, or currently have any capabilities to traverse, anything beyond the orbit of the highest satellites. Most of the cooperation between it and NASA, for the foreseeable future, will likely be the development of permanent communication channels—laying the cables so that, when a 2022 TTX scenario unfolds, there is minimal dawdling.

Even if these protocols have been prepared decades in advance, six months warning is still a nightmarish scenario. "It takes years to design a mission to deflect an asteroid," said Chodas. "Six months is simply not enough time" to deploy any effective mission, regardless of the type. "Getting a NED to the asteroid requires a warning time of at least five years, in my opinion. Maybe we can compress that a bit." In our alternate timeline, in February 2022, with a possible August impact looming, only two mitigation options existed: a last-ditch and likely futile effort by one or various nations to disrupt the asteroid with a nuke, or an evacuation of the future ground zero.

At this stage, it wasn't certain that America was the destination of the asteroid: the dungeon masters explained that there was a 19 percent chance that impact damage would be inflicted upon the nation. That uncertainty was grim by itself, and there was no clear way for anyone to do anything to protect themselves. But it could have been worse, globally speaking. What if the impact ended up happening somewhere else in the world? Imagine the geopolitical repercussions of an asteroid slamming into the Ukrainian-Russian border region, or the dividing line between North and South Korea. Or imagine the toll exacted by an impact on the Indian subcontinent.

In April 2015, the latter scenario was gamed out during an exercise[18] held in Frascati, Italy. The planetary defense war game of 2015 were more improvised and casual than the 2022 version, but it provides a fascinating study into how the fears spurred by the coming asteroid might generate alarming situations that compound the horrors of the impact itself. A group of conference attendees, taking on different roles—political leaders of directly affected nations, leaders of indirectly affected nations, residents of various countries, and the media—had to throw out suggestions about what they would do if they had seven years to mitigate an asteroid impact. It was a rollercoaster.

Initially, life went on as usual. Governments largely ignored the possible impact until its probability rose to disquieting levels. Parts of the western Pacific Ocean and Asia looked vulnerable. Nuclear-armed Pakistan raised concerns that nuclear-armed India would try and outdo them in the race to use NEDs to stop the asteroid. Iran assumed the asteroid was a hoax, and suspected it was a ploy to act against them. These nations were concerned that those with more advanced space programs would do nothing to stop the impact, since they would not be directly affected. Nobody knew who was supposed to do what, or who was to be trusted. If the impact did happen, who would help fund the rebuilding efforts? There was worry that corruption among public officials in certain countries would stop aid getting to the right people.

Eventually, rumors surfaced that nations were covertly planning unilateral action to launch nuclear warheads at the asteroid. The public began to get extremely worried. "We're all going to die" became a common refrain on Twitter. Misinformation spread everywhere. Impact zone insurance scams robbed millions of their life savings. Financial markets were thrown into turmoil. America wasn't trusted. China pondered teaming up with India to launch a mission. Pacific Rim nations, fearing a lethal tsunami, weren't sure if they should be evacuating.

By the time an impact was certain, the impact corridor included the Philippines, Vietnam, Laos, Thailand, Myanmar, Bangladesh, India, Afghanistan, Pakistan, Iran, Iraq, and Turkey. No matter where it hit, many would die, and there would be political, social, and economic havoc. The panic was self-fulfilling: the value of the Philippine peso dropped by 50 percent, and other nations stopped paying their international loans. Nationalism and xenophobia spiked. Some people wondered if the impact was god's incontrovertible will.

Some nations became NED hardliners, while Pacific Rim countries were against using them. Archrivals Pakistan and India tried to work together on a nuclear-armed spacecraft mission, but India withdrew when a DART-like nonnuclear impactor was suggested. The extreme tension led to acute geopolitical fears. What if China took advantage of the chaos to invade Taiwan? What if Iran's regime developed nuclear weapons under the guise of planetary protection?

Eventually, just three years prior to impact, multiple spacefaring nations agreed to launch multiple kinetic impactors at the asteroid—now thought to be of country-crushing size—to try and deflect it. Six were launched, and four were needed to succeed. India dissented, saying that it would probably send its own nuclear-armed mission regardless. It would also strengthen its borders to prevent a mass exodus of refugees coming in from Bangladesh or Pakistan. But then China announced that it, too, would launch a nuclear device to save the world. Both countries oscillated between deploying nukes, evacuating people, or doing nothing and letting the planet take a hit which, optimistically, wouldn't permanently scar their own countries. America offered support on the auxiliary NED option in case the kinetic impactors failed.

They did. About eighteen months prior to impact, the impactors shattered the asteroid into two pieces, with the smaller piece—about twice the size of Tunguska—still on course for Earth. As the nuclear option became the last hope, nations became intensely suspicious of each other, in some cases threatening military responses to thwart either a rival's nuclear buildup, or an accidental nuclear explosion on the ground in the region. China decided that they could mostly absorb a two-Tunguska impact, and protested other countries' NED mitigation plans.[19]

Just a week before impact, the impact location was confirmed: close to the city of Dhaka, the capital of Bangladesh. Up to 27 million people could be directly affected by the blast, equivalent to 18 million tons of TNT. Lewis was at the 2015 conference, and he was aghast that some of those playing the role of the neighboring nations closed their borders. Being a self-professed wise guy, and a retired naval officer, he made a bold suggestion. “Alright, I’m going to take what’s left of my meager armed forces, I’m going to pick a spot on the border, and I’m going to attack,” he said. “Why would you do that?” they asked him. “I need to save as many of my people as possible. I’m going to pick a spot on the border, I’m going to attack, open it up, and try to get as many of my people across as possible before the impact. How about that?”

The response was mixed. Some looked horrified. Others said that a pre-impact conflict was justified for humanitarian reasons. Federica Spoto, who researches asteroid dynamics at the Harvard-Smithsonian Center for Astrophysics in Massachusetts, remembered that exercise well. “For populations that have hated each other for centuries, you can’t just tell them to move and go with their neighbor,” she told me. The border war moment “was crazy,” but not ridiculous. “It’s chaos. I think it’s not so far from the reality, maybe.”

But in our alternate timeline of 2022, these nations remained unscathed. Instead, it was looking more and more like America was the target. And by June 2022, just two months before impact, and with no way to prevent it, additional observations pushed the odds to 100 percent. On August 16, an asteroid was going to hit North Carolina.[20]

At this point, Lewis explained, some people would just flee the state, maybe even the surrounding states. The turmoil would cause untold socioeconomic damage. Without a more specific impact location, the government wouldn’t be sure if they should proceed with an evacuation, or even how to proceed with an evacuation on such a large scale. There is no playbook for this cataclysm.

Vernon and his team of state and local officials, while swatting away and neutering as many falsehoods about the asteroid as possible, also found themselves overloaded with information gushing from the official NASA pipeline. Complex scientific facts were blasted into their ears and eyeballs. "This is not going to work," he said—they needed one-page summaries that anyone could understand. What, where, when, who, why. "We don't need pages of technical details. That data works for them. But when it goes public, you have to plan for that." They, and disaster management overseers FEMA, require simplicity to cut through the noise, with the uncertainties clearly explained. They don't need a scientist running into their offices with an unstable pile of papers, *Armageddon*-style. "That's called disorganized chaos," he said. "We try to organize the chaos as much as we can."

At this late stage, the hope was that someone trusted by the community—and the country, if possible—would become the eloquent, human face of crisis management. "You need, like, a Bill Nye the Science Guy," said Vernon. That could be the president, depending on who that is at the time. Maybe it would be someone at NASA, or FEMA. "Someone who's good with people." Perhaps a local mayor, or city manager. "Me? I don't know," he said.

With just two months until impact, everything was awful and unclear—not just locally, or nationally, but internationally. This asteroid would consume everyone's entire life until it hit. Who would coordinate the recovery operation? Who would coordinate the overseas response? Who would keep the bad actors out? And when should NASA hand over the lead agency role to FEMA? Lewis, a man of both worlds, did not know. Nobody knew. All they knew was that North Carolina's days were numbered.

You might think it strange that astronomers cannot pin down the exact location of an asteroid that is ringing the planet's doorbell. Optical astronomy has its limits, and not just because it can provide only a crude estimate of an asteroid's size through its reflected sunlight. A speck of light moving through space provides a way for scientists to work out how an object orbits the Sun. But space is huge. Using that visual track across the sky to work out the precise location an asteroid will touch grass is impossible. A reconnaissance mission, like a version of OSIRIS-REx, could determine an asteroid's path through space with unparalleled fidelity. But you can't build and launch one of those in six months.

That leaves one thing left in the toolbox: radar.

Ed Rivera-Valentín, a member of the planetary radar science group at Puerto Rico's Arecibo Observatory, filled me in on the details. Radar in the military and radar in astronomy work pretty much the same way: you beam out invisible radio waves, and if they hit something, they bounce back. Do that frequently and hastily enough, and you can identify an object, determine its size and shape, and trace its movements with a remarkable degree of precision. It even works on an asteroid zipping through space at tens of thousands of miles per hour, with its ascertained size and trajectory accurate to the nearest few feet, sometimes even inches. That sounds like sorcery to me, I said—it's absurd. "It is! It's ridiculously absurd," Rivera-Valentín exclaimed.

Radar observatories aren't all made equal. A larger dish can gobble up more incoming radio waves. "The bigger your dish, the smaller the things you can see," they explained. And to ping more distant objects, you need heftier batteries. "The more power you're emitting, the farther away you can see." Radar sounds great—so why bother using optical observatories at all, after they initially find those streaks of light?

It is, of course, because there is a massive catch. Light can travel through space for eons without exhausting itself. Radar can only bounce radio waves off asteroids that are relatively close to the planet: several million miles or so away. That sounds far, but not when those objects can cover those distances in just a few days. That is why the exact impact location of Winston-Salem, North Carolina, was revealed just six days before impact—nowhere near enough time to get everyone out of the way. This radar lock also confirmed 2022 TTX's size: 230 feet, a little larger than Tunguska, and sufficient to destroy the entire city. At 2:02 p.m. on August 16, it would hit the atmosphere at 35,000 miles per hour at a steep angle, and likely create an airburst that would cause widespread devastation.[21]

State and local officials were used to uncertainty in their jobs. Hurricanes, for example, can take many paths, and meteorologists use a cone of possibility to show the multiple paths a storm may take prior to landfall. But these possible routes are often close to one another, and a hurricane is unlikely to veer wildly off course moments before it meets the coast. That you can only know which segment of a wide region is going to experience a catastrophic asteroid impact less than a week before it happens was baffling to several players of the exercise. The explanation, that radar doesn't work until the asteroid gets close to Earth, proved to be both clear and alarming.

Six days to go. More than a quarter of a million people to try and shield. Evacuation was the only option—sheltering in place would have been largely ineffective—but as with any incoming storm, no evacuation would be entirely successful. FEMA, which cannot order evacuations, would work closely with state and local officials, who would do their best to shepherd as many people as possible out of harm's way. But 250,000 people is far too many to move in under a week.

And, as with any oncoming tempest, some people wouldn't want to leave. They don't want to leave their property unprotected. Some may not believe the threat exists. "Some people flip us off, tell us to get lost, say they're not leaving," said Lewis. Pets are not always welcome in shelters. While some animals are let loose to fend for themselves, most won't be left behind by their owners, because pets aren't like family members—they *are* family.[22] In fact, staying behind to do something, anything at all, to protect their four-legged friends is the number-one reason why people don't evacuate, Vernon told me.

You would also have to switch off a city's infrastructure to limit the impact's damage. If you could, you would shut down sewer lines, water lines, and power plants. Doing so preemptively may encourage more stubborn people to evacuate, said Vernon. Resources vital to the recovery effort, from fire trucks to medical equipment, would be sped out of the city, along with hospital inpatients, some of whom would inevitably die in transit.

As part of the war game, NASA's Asteroid Threat Assessment Project, based at the Ames Research Center at Moffett Field in Mountain View, California, provided maps estimating the damage. Lorien Wheeler and Michael Aftosmis, two members of ATAP, told me that these estimates are not easy to make. They use supercomputers to consume all the available data about a hypothetical incoming asteroid, run it through uncountable simulations, and expectorate several possible outcomes. Regular computing power won't cut it: you're solving a hugely detailed physics problem in fine detail on the scale of a human population. That's both complicated and expensive. At Ames, Aftosmis said, "We host some of NASA's largest supercomputers, and so we run those simulations. They're supercomputer-size problems."

These simulations are overseen by ATAP members with different métiers, from atmospheric entry experts to blast damage specialists. And none of them speak with certainty. You can run a million simulations a day on cutting-edge gear, but without much real-world information to fine tune these virtual ventures, damage estimates are, well, more like guesstimates.

"Sometimes we know about the damage, but nothing about what came in," Wheeler told me. "Sometimes we'll have observations of smaller objects that come in, they'll make a flare, a light curve, but they're not enough to cause damage. Then we have data on very few actual asteroids before they enter," including Chelyabinsk. "But we don't have anything that connects all those pieces. That's why we build our probabilistic models that can run millions and millions of cases that look at all these maybes"—because some guidance is far better than nothing at all.

The Cold War helps them out a little: plenty of information on the physics and consequences of extremely large blasts comes from nuclear weapons research. "And those are key data points for getting ground truth," Aftosmis told me.

There are some differences. Asteroids don't spread spaceborne radiation everywhere, and those the size of 2022 TTX wouldn't produce a prodigious fireball. But the airburst's blast wave would slam into the ground and cause the ambient air pressure to shoot up. The parts of a city experiencing the highest overpressures would likely suffer the worst damage. What, I wonder, would that feel like? "If you were jumping off a bridge and slamming into water," said Wheeler. "Water's not hard, necessarily, but if you hit it at velocity your body is . . . going to feel a physical force and not be well afterwards." Oof.

In a nutshell, ATAP's role "is really a game of trying to figure out what we know best," she said. "People making decisions really want a single answer. And we can't give them a single answer." Any city facing the prospect of asteroidal annihilation won't know exactly what will happen until it happens. But it certainly will be gut-wrenching. "If Tunguska happened over a highly populated area, it would be a very large disaster," said Wheeler.

And, at 2:02 p.m. eastern time on August 16, Winston-Salem met 2022 TTX, and both ceased to exist.

One cold, crisp day in London, I sat in a South Kensington pub with Gareth Collins, eating burgers, drinking cider, and discussing the best way to destroy a city. Luckily, nobody overheard us and reported us to the authorities.

Collins is an impacts expert at Imperial College London, my undergraduate alma mater. He shares with many of his colleagues, and me, a macabre fascination with asteroids hitting Earth—low-probability, high-consequence events. "It's weird to think about something that won't happen for 100 million years, but when it does happen it will kill everyone on the planet," he said, eating a chip.

It is tricky for most of us to picture the impact of a planet killer. A city killer isn't quite as difficult because we have all seen footage of nuclear bomb explosions. "I show my students a few videos of the effects of nuclear explosion tests," he said. Then he asks them to guess how far they need to be standing away from these fulminations to be fine. It's often farther than they think. That, in essence, is what a city killer asteroid is like.

Like the members of ATAP, Collins acknowledges the uncertainty about what damage a hypersonic midair explosion can do. "Generally speaking—deeper, stronger, steeper, faster: all worse," he quipped. In other words, dense, structurally sound asteroids that penetrate deeper into the atmosphere inflict more damage on the local area. But other key parts of an impact event that feel intuitive turn out to be mysteries. It isn't clear which would cause more damage, an airburst or a ground impact. We haven't experienced enough city killer impacts, or found enough of their craters, to solve that riddle.

All anyone can do is try. Collins, hoping to make impacts more accessible, so to speak, to nonacademics, was one of the creators of the website, Earth Impact Effects Program.[23] This site lets you build your own impactor—asteroid or comet—and throw it at the planet as fast as you like, into whatever you wish, and then it tells you what is likely to happen, from the damage it would do to various buildings to the violence that would befall humans. It is a darkly engrossing way to spend your lunchtime. How else do you think I worked out how to destroy Seattle?

The level of detail is astounding and unnerving. I casually lob a 500-foot iron-rich asteroid 45,000 miles per hour at a city built on a foundation of solid rock. The asteroid, the program tells me, breaks up in midair, creating not only an airburst but multiple ground impacts that hit so hard that any rock not vaporized immediately becomes molten, infilling a deep crater nearly two miles wide.

The unemotional tone of the program is unintentionally comedic. When it asked me how far from the impact I was standing, I put 300 feet—which might as well have been zero. "Your position was inside the transient crater and ejected upon impact." I became a piñata mid-smack, bursting into a thousand pieces while being propelled skyward. Fiddling with the values again, I find myself standing much farther away, but now without a mode of transportation: "Cars will be largely displaced and grossly distorted and will require rebuilding before use." You know what? I think it might be easier to get a new car. More finagling, and I find that my clothes explode into flames and my eardrums experience the equivalent of a fighter jet taking off right next to me.

All best guesses, said Collins—and generalizations. A city killer asteroid won't take out every person, and every building, in a city. Some may be shielded from the blast by a hill, or may simply be more resilient than other pieces of infrastructure. Others will topple like a house of cards. Perhaps the city remains largely intact, but the aftermath seals its fate. "Half the city could be on fire for days," he said. He looked at the pedestrians milling about outside the pub. "We'd have a new fire of London, and maybe that turns out to be the worst effect." Hospitals could be overwhelmed. There may be sewage all over the place. There might not be clean water or electricity for months, or years.

Programs like this can tell you how likely you are to perish or be maimed in various impact scenarios. But the implication of every single one of these simulations is identical. "The message, really, is that our existence is quite fragile," said Collins. I agreed with his assessment, and we ordered another round of drinks.

Eight miles above Winston-Salem, 2022 TTX exploded.[24]

In a heartbeat, the blast wave knocked over, eviscerated, or mauled most buildings within a 40-square-mile patch. Widespread infrastructural damage occurred across a 425-square-mile area, and windows imploded across a 2,600-square-mile expanse. Fires sprouted like weeds. Corpses littered the torn-up streets. Thousands of survivors were trapped between and beneath rubble. The first satellite images of the impact psychologically scarred the nation, a trauma that was compounded by footage from drones and first responders as they made their way toward ground zero. Some of the modeling suggested that "there'd be so much dust that you couldn't see for days," Vernon told me. "It wasn't *quite* like a nuclear bomb." But it wasn't that far off.

Disorder reigned. There was scattershot looting. Nobody was quite sure how to maintain order, or when it would be safe to allow people back to search the rubble of their homes, if ever. Not long after the effort to recover survivors began, a prominent purveyor of misinformation, being interviewed by less meticulous news outlets, claimed that the asteroid contained toxic materials from outer space, and anyone exposed would experience something like radiation sickness. And while America tended to its grave physical and psychological wounds, people began to ask: What happens next? The extent of the destruction meant that reconstruction would take decades. "Is it going to be worth rebuilding?" was a fair question to ask, Lewis told me. How about insurance? "'Sorry everyone, it's an act of god, we're writing off everything, goodbye.' Do we let them do that?"

Survivors might protest that nothing at all was done to stop the asteroid impact. Why not just launch every nuke we had atop any rocket powerful enough to carry them into deep space? Even if everyone said it would be for naught, we'd still have to try, right?

"I'm a firm believer in the law of unintended consequences," said Collins. If you try to disrupt or deflect an asteroid with NEDs, and it fails, you might get a fragmented cluster of smaller asteroids that strike not one country, but many, leaving behind multiple irradiated holes in the ground. Even improvising a DART-like kinetic impact mission at the last moment may knock an asteroid into a more densely populated city, state, or country. In some cases, doing nothing may not be inaction, but the least-uncertain, least-bad choice. "It does go against human nature, overcoming the human instinct to do something just because you can... [to] instead worry about getting people out of harm's way," said Collins.

Lewis agreed. "One of the decisions might be to take the hit," he said. Not one to dwell in darkness for too long, he then offered up a far more ideal scenario. "If you're going to tell me it's going to land in the Nevada desert, way out in the middle of nowhere—hey, you know what? For science, let's watch this joker! That'll wake everyone up for sure."

In fact, despite the seriousness of the subject matter, and coming away with similar lessons time and time again—that any impact scenario will involve a lot of uncertainty and dread—most involved in these asteroid-themed Dungeons & Dragons games remain upbeat. Like playing a horror video game, participating in these simulations of scary things leaping at you from the shadows doesn't involve any real harm. Participants get injury-free bursts of adrenaline, while contributing to a very genuine effort to save millions of real lives one day. And, frankly, if you can't find humor in even most the dismal moments of our lives, your mind will eventually unravel.

The dungeon masters even insert easter eggs into these games for people to discover, "little cookies" as Chodas described them. During a 2021 exercise,[25] in which an asteroid ended up hitting the border nexus of Germany, Austria, and Czech Republic, he conspired to have the impact location be the house of Detlef Koschny, a participant and the then-acting head of the European Space Agency's Planetary Defence Office. The local press took notice. "One of the biggest newspapers picked it up, put it on their front page," Koschny told me, laughing. "Now everybody here in the area knows I'm working on asteroids."

Both Lewis and Vernon epitomized that paradoxical good-natured enthusiasm more than anyone else involved during the 2022 planetary defense war game. It came as a pleasant surprise. These two people, who spend much of their time thinking about or planning to prevent some of the very worst disasters, should be miserable. But the unique nature of the threat of an asteroid impact is, I suspect, why they were so chipper. They both accept that most disasters are baroque; they feature multiple antagonists, both human and natural, combining to unleash a tsunami of problems that are impossible to completely prevent. Certainly, an asteroid impact would be an enormous challenge. "There's no Planet B. There's no starships to evacuate us. It's going to take all hands on deck," Lewis concluded. But he and everyone else playing that 2022 war game knew that cities like Winston-Salem don't have to die. If the world supports the right people, all future asteroid impacts can be prevented, and everybody gets to live—a thought that should put a smile on anyone's face.

And, while that war game was taking place, the right sorts of people were already being supported. The midflight DART mission was proof that half of our planetary defense system was under construction. But what of the other half? What about our ability to detect these asteroids, to find them before they find us? As that 2022 exercise demonstrated, it is perfectly possible that, with our current NEO-hunting apparatus, the nightmare scenario of a city killer appearing just a few years or months prior to impact can happen—a situation that would make DART and NEDs as useful as a chocolate teapot. How, then, do we end that nightmare? How do we push the months and years into decades, or centuries? That requires a special sort of observatory, one that can break the chains of gravity, one that can find even the stealthiest asteroids faster than anyone else.

What the world needs is NEO Surveyor.

A weirdly prescient billboard in New York City. *(Robin George Andrews)*

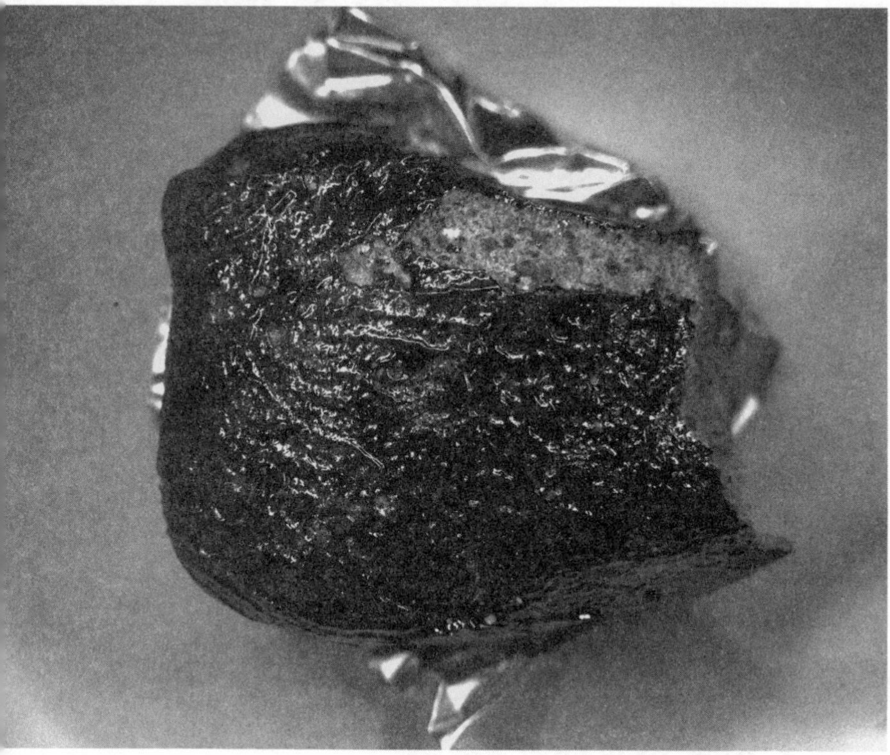

One of the Botswana meteorites found in the wildlife-packed Kalahari. *(Mohutsiwa Gabadirwe)*

Kagiso Kgetse, a wildlife officer, finds a small meteorite hidden in the Central Kalahari Game Reserve. *(Mohutsiwa Gabadirwe)*

A large chunk of the Winchcombe meteor crash lands and disintegrates on the Wilcocks' driveway. *(The Wilcock family)*

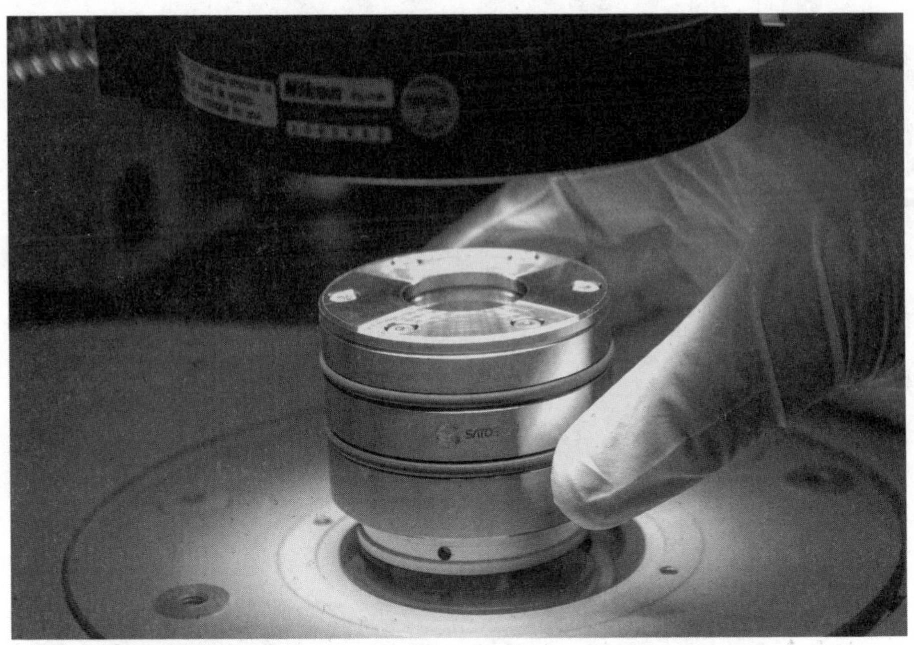

A Hayabusa2 sample canister containing fragments of the asteroid Ryugu is transferred from JAXA to NASA. *(NASA / Robert Markowitz)*

Arizona's Meteor Crater. *(Robin George Andrews)*

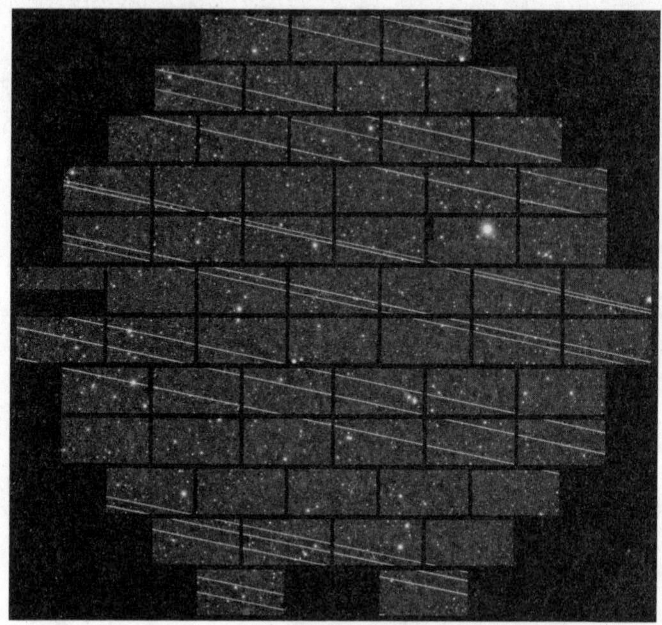

An image taken by the Blanco 4-meter Telescope at the Cerro Tololo Inter-American Observatory, featuring at least 19 obstructive Starlink streaks. *(NSF / Clara Martínez-Vázquez and Cliff Johnson / CTIO / AURA / DELVE)*

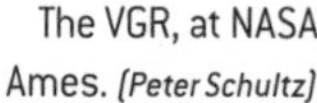

The VGR, at NASA Ames. *(Peter Schultz)*

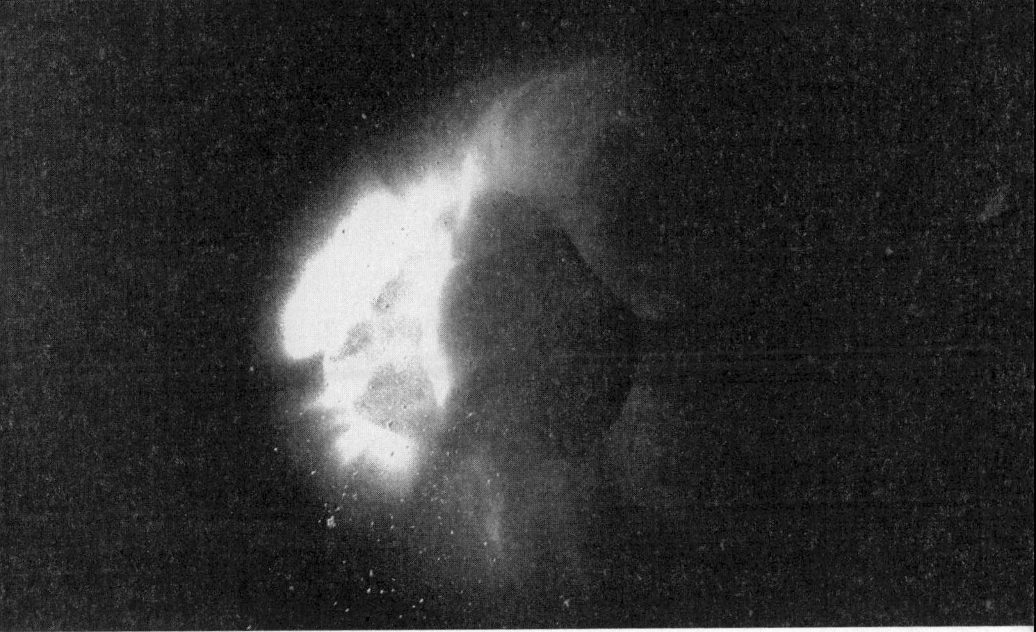

Shooting at an asteroid fragment with the VGR. *(Peter Schultz)*

Blasting out impact craters at the VGR. *(Peter Schultz)*

VIII

In the Shadow of the Sun

In 2019, Hollywood gave Amy Mainzer, an astronomer at the University of Arizona, a call. Director Adam McKay—best known for movies such as *The Big Short* and *Vice*—was making an altogether different flavor of film, one in which astronomers discover a planet killer comet heading for Earth. It would be a tragicomic drama. Leonardo DiCaprio and Jennifer Lawrence would be the astronomers. Meryl Streep would be the Trump-like president. And Mainzer would be the film's scientific advisor. It didn't take much consideration from Mainzer to agree to help.

How was it? "A ton of fun," Mainzer told me. Of course it was—even though the movie wasn't actually about an impact threat. Sure, there's a comet. But it's a plot device to spread a more universal message about our current misinformation-plagued era: that no matter the disaster, powerful people will tell the masses not to listen to scientists, even if they are screaming bloody murder, until it's too late. A big clue is in the title: *Don't Look Up*.

One crucial scene features DiCaprio as an astronomer from Michigan State University berating cable news hosts who play up the danger for laughs, claiming that their audience doesn't want to hear doom and gloom. During his indignant, on-air outburst, he screams: "If we can't all agree at the bare minimum that a giant comet the size of Mount Everest, hurtling its way toward planet Earth, is not a fucking good thing, then what the hell happened to us?"[1] To scientists, the scene proved simultaneously gratifying and mortifying. "When Leo finally goes off on his big tear about this," Mainzer said, "I told him, we've been through this, we've been through this many times actually. That's how a lot of us feel, I think. We have to be able to talk to each other."

When Mainzer first saw a draft of the script, nobody had heard of COVID-19. But then everyone was aware of it. The ensuing coronavirus pandemic caused more than dread and death—it produced a torrent of bad-faith actors who flooded the mediascape with all manner of falsehoods. Of course, it was nothing new for people to ignore science, or even intentionally obfuscate facts and downplay the seriousness of something that affects everyone negatively. The issue of climate change provides many examples. But the COVID-19 pandemic opened portals to many disturbing dimensions, such as anti-vaccine cultists from all backgrounds who compared epidemiologists and virologists to Nazis.

Speaking from Tucson, Arizona, Mainzer's face crumpled while thinking about the worst days of the pandemic. The rejection of basic scientific principles alone "was really frightening," she said. "Just because you can't see something with your eyes, doesn't mean it's not there, whether it's a carbon dioxide molecule, a methane molecule, a virus, or an asteroid that you have to look at through a telescope."

After perusing the movie's draft script, she had one overarching piece of advice for McKay. "You've just got to make it weirder and weirder," she said. McKay did—and by the time *Don't Look Up* was released in December 2021, even some of its most outlandish plot points felt less shocking than they should have been. "This fundamentally is a movie about the importance of science," she told me. And in this case, the science was about saving everyone from certain doom.

Mainzer is no stranger to that idea. She is the head of a mission named Near-Earth Object Surveyor, or NEO Surveyor, the asteroid-hunting observatory to rule them all. When built and launched into space, it will turn its gaze toward the Sun's almighty glare in search of the sort of city killer asteroids that are especially difficult to find with almost any other telescope. Remember those 15,000 NEOs out there—the ones about the size of Dimorphos—that we have yet to discover? In just ten years or so, NEO Surveyor will be able to locate 90 percent of them, allowing humanity to definitively answer an existential question: are we in danger of being hit by a killer asteroid in the next century, or not? If the answer is affirmative, then we will have time to do something about it with our DARTs and NEDs. If it's negative, we can all breathe a massive sigh of relief for at least a lifetime.

Without NEO Surveyor, though, the picture will remain unnervingly incomplete. Kinetic impactors cannot hunt for killer asteroids if they can't see them coming. So why isn't this technological sentinel already operating above all our heads? Planetary defenders almost universally support Mainzer and her mission. They all desperately want everyone to look up. The problem, it turns out, was NASA itself: the very same organization that had set up the Planetary Defense Coordination Office had, for several decades, been unsure if its top priority should be defending the planet.

Mainzer always had an eye for the invisible—at least, things invisible to the human eye. She is a longtime fan of the infrared part of the electromagnetic spectrum, the red-hued band, the bit that comes up whenever people talk about heat signatures. If something is warm, you can use an infrared sensor to detect it. The same applies to asteroids. Out in the cold expanse of space, they can get almost as chilly as the nothingness of space itself. But bathe them in a bit of sunlight and they get a little toasty. An infrared sensor could probably pick them up, Mainzer thought—so she joined the Wide-field Infrared Survey Explorer, or WISE, mission team.

WISE, a satellite launched in 2009, uses its infrared eye to look for anything it can detect in the night sky, from ultracool stars to the brightest galaxies.[2] Mainzer became the principal investigator of NEOWISE, the asteroid-hunting part of the mission. Before the WISE satellite went into an extended hibernation, it managed to find 34,000 newly discovered minor planets—asteroids, comets, other big things that aren't quite moons or planets—all over the place. Although it was recently reactivated, the satellite's days are numbered: its orbit is decaying, and it will soon burn up in Earth's atmosphere.

NEOWISE did a respectable asteroid-hunting job. But from a planetary defense standpoint, it mostly found the sort we largely aren't concerned with—those in the belt between Mars and Jupiter, and only a few that have escaped and are now zipping about close to Earth.

"This telescope is not designed to look for near-Earth asteroids, it's just not," Mainzer said. What was needed, she thought, was a mission specifically made to hunt for asteroids. And it would still do that in infrared, because infrared is the Sherlock Holmes to everyone else's low-rent detective: it finds its marks better than anything else on the spectrum.

Visible light is fine for the initial discovery. But a chalklike asteroid that is small and reflective looks the same as a charcoal-esque asteroid that is big yet dull. They reflect the same amount of light back to Earth.[3] Is it a country crusher or a city killer? If we rely on visible light, we may not know until it gets much closer and radar locks onto it. Infrared is different. A smaller asteroid emits less infrared radiation than its bigger brethren. Infrared does not depend on the nature of an object's light-reflecting coating. Small rocks are like lit matches, and bigger beasts are bright like bonfires—no exceptions. "That's why it's a pretty handy tool," said Mainzer. The same principle conveniently applies to comets, which are infamously dim until they suddenly burst into life not far from Earth.

Visible light astronomy has other problems. The stars may be beautiful, but to asteroid hunters, they are photobombers. Infrared astronomy filters out plenty of these shiny intruders. In theory, a dedicated asteroid-hunting mission with infrared eyes could solve all of today's problems with NEO detection. The current crop of largely ground-based optical observatories is doing its best. But an infrared endeavor would dramatically accelerate the discovery rate.

Such a mission, however, comes with a bigger price tag. Ideally, you would want to launch an infrared telescope into space to avoid any Earth-based electromagnetic and atmospheric interference, and to give it the widest possible field of view. You would also need to make sure the spacecraft itself was one of the coldest objects in the universe—using chemical coolants, or by giving it a cleverly designed deep-space parasol. That way, the heat from the spacecraft's own instruments, or any heat generated by the absorption of sunlight, wouldn't interfere with its ultrasensitive infrared sensors. All that adds up. But hey, what price could be too high for the vanguard of the detection-based half of Earth's planetary defense system?

Thus, the idea for the Near-Earth Object Surveyor was born—and Mainzer would lead the charge. If it worked as planned, she said, 90 percent of all asteroids 460 feet and larger would be found within just a decade. That sounds ambitious, but the science is sound. "Asteroids are scattered all over Earth's orbit. They're all over the place," she said, showing me a plot of all the known NEOs. It looked like a wild Etch A Sketch drawing. There are lots of potentially hazardous objects yet to be found, but if we can spy all these NEOs already using visual light, imagine what will be revealed with infrared radiation. "At the end of the day, it's finding rocks in space. That's its sole goal. We're not trying to do anything else. It's looking for asteroids and comets," Mainzer said. "It's a pretty predictable physics problem."

Another key to NEO Surveyor's almost guaranteed success is its position in space, far from home. Many ground-based surveys wish to find asteroids lurking in the part of space interior to Earth's orbit, which just so happens to be an area flooded by the Sun's glare. You cannot point terrestrial observatories directly at the space around the Sun; even large asteroids will be obscured by the fountain of light. All anybody can do is aim their telescopes very close to the Sun during twilight, just after sunset or immediately preceding sunrise—giving them just minutes of time to peer through a thick atmospheric haze and find asteroids camouflaged by honeyed rays of gold.

NEO Surveyor, alone in space, can stare nearly directly at the Sun and, with its infrared sensors, find the asteroids almost nobody else can. And it may find something bigger than a city killer. In 2022, a team led by Scott Sheppard, an astronomer at the Carnegie Institution for Science in Washington, DC, found a mile-long asteroid previously concealed by our local star's infernal light. "That is what we call a planet killer," Sheppard told me.[4] He added: "We expect that most of these are known.... It's rare for these to still be out there and [for us to] not know about them. And the reason we don't know about them is because you have to look towards the Sun to find these objects." City killers remain the primary threat because of how numerous they are. But the implication of Sheppard's discovery was clear: a handful of planet killers are still out there in the darkness.

The object, dubbed 2022 AP7, was found by a ground-based observatory in Chile, one investigating dark energy, which is thought to be driving the accelerating expansion of the universe. Similar NEO discoveries will happen. But if we really want to find them all, and find them fast, we need NEO Surveyor—the science is inarguable. "My big hope for this is we don't find anything heading our way," said Mainzer.

"That would be the best answer. That would be great." But it's important to know for sure, not just cross our fingers.

I spoke with Mainzer in 2022. That year, DART was heading toward its target, while NEO Surveyor was languishing in development hell, and only existed on paper. What in the name of Dimorphos was going on?

The concept for NEO Surveyor dates to the early 2000s. Back then, it was known as NEOCam, but it was essentially the same thing: an infrared near-Earth asteroid hunter. And like many robotic missions hoping to get NASA's assent and become real spacecraft someday, it had to enter a competition against other mission designs and emerge victorious. It never did.

The cheapest competition category, the Discovery class, runs once every few years or so. Discovery missions must have a very specific, limited design, with a cost cap of around $600 million, all-in. A good example of this was the InSight mission, a solar-powered lander whose sole job (not that it was easy) was to image the geologic insides of Mars. The next class up, New Frontiers, with its $1 billion spending cap, allows for a more multipurpose mission design and can include more advanced technological capabilities. Dragonfly, a nuclear-powered quadcopter that will head to the Saturnian moon of Titan, was a recent New Frontiers winner. In both cases, teams led by principal investigators pitch designs, and panels made of experts at and outside of NASA judge them and eventually pick winners.[5] There are rarely bad ideas, which means getting chosen is remarkably difficult.

Those that lose out can tweak their design and reenter in subsequent competition rounds, sometimes with some consolatory funding to help them out. NEOCam, a relatively simple design with a very narrow purpose, was proposed several times through the Discovery competition. In the mid-2000s round, the mission concept was viewed positively, but scientists had yet to invent long-lasting, high-resolution, and extremely cold (and therefore sufficiently sensitive) infrared detectors to go with it. Come back later, NASA said. When the technology had matured, NEOCam was reentered into Discovery twice more in the 2010s but lost both times[6]—although NASA did bequeath Mainzer's mission concept with some funds to help develop its infrared detectors.

"Eventually, the project got moved into the newly created Planetary Defense Coordination Office," said Mainzer. In 2019, NASA's leadership—those above the PDCO—decided that a spacecraft designed to prevent the massive loss of life on Earth meant that it shouldn't compete for attention among planetary exploration missions. A year after DART was officially approved,[7] NEOCam became a "directed mission," which meant it would not be subjected to another competition gauntlet and would instead be given its own funding stream.[8]

The Discovery competition victors have always been worthy selections. But pitting a planetary defense spacecraft against pure science missions was a major mistake. If DART had also been forced to do the same, it may never have seen the light of day. At the time of NEOCam's transferal to the PDCO, Thomas Zurbuchen, who at the time was the associate administrator for the Science Mission Directorate—the part of NASA that steers its scientific investigation of the cosmos—said[9] that science had been his metric for approving new missions. Although he held this post at NASA only from 2016 through 2022, this line of thinking was likely shared by his peers and predecessors, which meant that, in competition, planetary defense was always at a disadvantage.

"NASA insisted that we change the name, so we changed the name" from NEOCam to NEO Surveyor. Why? It's the same mission, right? Mainzer nodded, then said: "You should ask them. What's in a name, right?" From 2003 up until that point, Mainzer had been working as a scientist at NASA's Jet Propulsion Laboratory. But in 2019, as her mission escaped its Sisyphean loop, she left the laboratory,[10] moving to the University of Arizona while retaining her leadership position with NEO Surveyor.

NEO Surveyor's fate was looking promising. The instrument design was bleeding edge, on point, and streamlined. There would be no need for chemical coolants that would eventually run dry. "It's got a big shiny sunshade on one side of it, that'll block the sun," said Mainzer. "We paint the back surface really, really black, with a pitch-black paint, and that functions as a very efficient radiator." The funding appeared to be there. The team was ready to get on with construction and prepare for a launch in the not-too-distant future, perhaps 2025.

And that's when things got weird.

NASA flight projects have design phases,[11] labeled alphabetically. Phases A and B are largely conceptual: tweaking mission design, planning out the entire project, making sure the technology used will work in space, all that jazz. If NASA leadership allows the mission to make it past Phase B, they give the mission a formal budget and the team can begin building the thing.[12]

NEOCam had been refined through multiple rounds of Discovery. But inexplicably, it remained stuck in Phase A—the bit designed to establish a baseline concept—until the summer of 2021, at which time it moved[13] into Phase B, the technology and engineering prototyping stage of the process. It would not go on to pass its Phase B review until December 6, 2022.[14]

It was then, just before Christmas of that year, that I finally heard back from Mainzer. I had been told by a dozen different people that she was swamped and couldn't talk. Now that that major milestone had been passed, she could finally chat—and she sounded not so much thrilled, but relieved. They were in Phase C,[15] the stage at which NEO Surveyor's final design could come together and in which hardware and software could be crafted. Next up, Phase D: assembly, testing, and ultimately launching the spacecraft.

Mainzer emphasized that, like all mission concepts, hers needed to be thoroughly reviewed by independent experts. "Space is a very unforgiving business. A lot of things can go wrong. So it really has to work," she told me. Finally, after everyone agreed that it would in fact work, NASA officially committed itself to the project, offering up $1.2 billion as a budget and promising a launch no later than June 2028,[16] though ideally it would fly in the mid-2020s. The curse appeared to be broken.

Or so they thought. "We had a big budget cut this year, which was not what we were expecting," said Mainzer. That $1.2 billion was to be split over ten years, with the yearly budget varying depending on the project's needs. In 2023, the mission required $170 million. As with all federal government funding requests, this needed to be approved by Congress. Historically, NASA got what it asked for in the planetary defense realm. But for some reason, just before NASA gave the NEO Surveyor team the go-ahead to start piecing the mission together, NASA's senior leadership asked Congress for just $40 million, rather than $170 million. "That . . . is a pretty big difference," Mainzer said, as dry as a desert. That plot twist came with no explanation from NASA. Mainzer was, to put it mildly, frustrated and confused.

So was I. This equation seemed both simple (we need X amount to fund the eyes of the planetary defense effort) and inexplicable (NASA doesn't want to give their own mission X amount). So I called Casey Dreier, chief of space policy at The Planetary Society—a space exploration advocacy group—to help me out. Dreier is a numbers guy, the person who translates budgetary black magic into colloquial language. He'll explain to me how this makes sense, I thought.

This decision made no sense whatsoever, he told me. Oh, okay then.

One reading of the scant statements put out by NASA was that other major missions needed more money than they initially thought,[17] leading to NEO Surveyor—again, a mission designed to identify killer asteroids—being stripped of funds it was supposedly guaranteed.

What mission could be more vital than NEO Surveyor? One possibility was the NASA-led Mars Sample Return, a multi-decadal, multipart, multiagency attempt to get fresh and possibly ancient life-containing rock samples from Mars back to Earth. It is a bit like a Rube Goldberg machine, a complex series of launches, landings, and robots driving around the Red Planet that all must go exactly to plan or the entire mission fails. Although worthwhile—it is, after all, trying to determine if there was ever ancient life on our planetary neighbor—the mission has been increasingly plagued by delays and budgetary issues.[18] It may yet succeed, but it may also inadvertently take down several other missions. Another accidental antagonist may have been Psyche, a deep space mission featuring a robot designed to pursue and study a quirky, metal-rich asteroid, one that may be the exposed heart of a planet smashed to bits eons ago. It launched in October 2023, but its flight was delayed by a year and ran over budget.[19]

But when the budget cut to NEO Surveyor was announced, no clear explanation was given, so all we had were educated guesses. When I inquired, a NASA spokesperson had this to say, via email: "The $40m request you cite is from the FY23 President's Budget Request, which was drafted almost a year before NEO Surveyor completed its Preliminary Design Review and was then confirmed in Fall 2022. The FY24 President's Budget Request supports that development schedule, and the mission is on track for launch." True, NEO Surveyor had not yet been officially confirmed as going ahead when the budget cut was announced. But it was already a directed mission whose budgetary requests were being relayed by the NEO Surveyor team to NASA's leadership. Cutting the budget for your own mission, one barreling toward such a major milestone, is just shooting yourself in the foot.

"They just ran out of money, it seems like," said Dreier. He wondered why NASA simply didn't request more funding from Congress, which he strongly suspected they would have readily approved. It was unclear. "There's never been a good answer. They've not bothered to give a good answer to anybody, which means it's a bad answer," he told me. At this stage, getting a transparent explanation won't help anyone.

“The problem here is that the damage has already been done,” said Dreier. Ultimately, Congress gave the mission $90 million for 2023—$50 million more than NASA requested, but far short of the $170 million needed. A mid-2020s launch date was now impossible; that budget cut led to staffing layoffs, delays in obtaining vital hardware, and pushing NEO Surveyor into an era of high inflation and possible recession.[20]

The project “is a lot more expensive than it used to be,” said Mainzer. “All these delays have cost a lot of money.” Fortunately, having reached Phase C, NEO Surveyor is almost certainly safe from cancellation. But that it is being subjected to so many needless endurance tests underscores a concerning fact of life: no matter how you wish to save the world—whether it’s from a lethal virus, climate change, or potentially killer asteroids—you will always have to fight a lumbering beast named bureaucracy.

NEO Surveyor’s woes also echo another uncomfortable truth. America may finally be all-in on planetary defense, but the amount of money it gives to the effort is almost laughable.[21] “In the early 2000s, we’re talking single-digit millions of dollars,” said Dreier. “At the time, that was less than the travel budget for NASA staff. They just didn’t have a program to speak of.” In 2005, Congress told NASA that the space agency was legally required to find 90 percent of all 460-foot NEOs by the end of 2020,[22] a deadline it comprehensively failed to meet. That was in part because, from 2005 to 2010, NASA still had just $4 million per year—0.02 percent of their total budget—earmarked for its asteroid-hunting surveys.[23]

In 2010, the Obama administration proposed a fivefold jump in funding for NASA’s NEO observation program. By 2019, with the PDCO now established, DART being built, and NEO Surveyor rescued from competitive hell, planetary defense had $150 million per year to spend.[24] That sounds like a lot. But it’s only 0.015 percent of the US military budget, which is close to $1 trillion.[25]

Planetary defense is one of the very few issues which, as Dreier points out, enjoys strong bipartisan support in the extremely polarized, typically dysfunctional US Congress. The American public also shares the belief that asteroids should not hit the planet. In 2018, the Pew Research Center conducted a survey[26] in which American adults were asked what NASA's priorities should be: 62 percent of respondents said that the monitoring of potentially hazardous asteroids should be a top priority, with another 29 percent saying it is important but should be a lower priority than other things. Only monitoring key parts of the climate sits above it, with 63 percent of respondents saying that that should be a top priority. Just 13 percent said that sending astronauts to the Moon—something NASA is throwing billions behind with its Artemis program—should be a top priority. (Incidentally, 9 percent said that monitoring possibly deadly asteroids either wasn't that important or shouldn't be done at all. Who were those people?)

Planetary defense is hard to dislike. "It's NASA itself that has been hyper-reluctant to aggressively invest in this," said Dreier. Even after that landmark 2005 congressional mandate, NASA asked for just $5 million for their NEO observation budget. "That's not a serious request. If NASA would have requested more money, Congress would have given it to them, no question."

Despite proportionally huge funding increases to the cause, NEO Surveyor's ongoing troubles show that planetary defense funding still isn't a given. "It's clearly still very vulnerable to expediency for other project priorities that NASA has, even if those project priorities are not saving the entire human race from extinction," Dreier said. The symbolism of Americans in space and on other worlds is undoubtedly strong, and the science they can conduct is also welcome. But working to save the world we already have is, just maybe, a tad more urgent.

Perhaps by the time you read this, NEO Surveyor will be about to leave the planet behind. Maybe it will already be up there, finding rocks in space we can't otherwise see. It is a mission whose success will let us know how safe, or in danger, we all may be. It will give us time to defend ourselves if we are under threat. And no matter the outcome, I suspect that Mainzer will have retained the same humility she exhibited throughout our conversation. "We know where a lot of galaxies are that are a zillion light years away, but we don't quite understand what's in our backyard, you know?" she told me. "There are a lot of problems out there. And this is one that I thought, you know, I could make a dent in."

Every single person I spoke to for this book told me that NEO Surveyor is as essential to our long-term protection as DART. "We can't deflect what we don't know is there," said Richard Binzel, the planetary scientist who invented the Torino Scale. "Luck is not a plan." He does not sit awake at night, fretting about all the asteroids we have yet to find. But he recoils at the thought that we may not do anything about those hidden hazards.[27] "The odds are in our favor that we're not in imminent danger. But every year you delay is one less year you have to deal with any object that is in fact a serious threat." All we have is time, he said. We shouldn't waste it like this. "This is a game of cosmic chicken. Who is going to blink first?"

And if NEO Surveyor does find that an asteroid is on its way toward us? "Then," said Binzel, "this would perhaps be the most important contribution that science can make to humanity."

Perhaps we do, despite everything, live in the best possible timeline—because NEO Surveyor will not be alone. A second next-generation observatory, one also capable of finding faint asteroids sneaking about, is going to join Mainzer's planetary-defending fight—and it will be on the verge of being fully operational by the time you pick up this book.

The namesake of the Vera C. Rubin Observatory is the late, great astronomer who revealed[28] that the universe has a ghostly glue keeping stars and galaxies more tightly bound than can be exclusively explained by gravity. That adhesive, dubbed dark matter, has not yet been identified by any conventional scientific means. The hope is that the new observatory will bring us closer to revelation. At the time of writing, engineers atop Chile's Cerro Pachón ridge are working on the finishing touches for the world's most advanced digital camera,[29] one that will spend at least a decade watching the entire night sky with its gigantic eye, soaking up the visible light (as well as some infrared and ultraviolet light) of every single object up there that emits or reflects it—an assignment known as the Legacy Survey of Space and Time, or LSST.

The Rubin Observatory is the magnum opus of optical astronomy. You may recall that telescopes can either see lots of the sky with a wide field of view, or they can have a big mirror to see the light from very faint and distant entities. "Rubin is the first observatory that has both: a large mirror, and a large field of view," Željko Ivezić, the director of the Rubin construction project at the University of Washington, told me. And it will leave other observatories in the dust. "In about ten years, we will have observed every point on the sky about a thousand times." Other telescopes would take 1,000 years. "It's impossible to do what Vera Rubin will do in ten years with any other telescope."

Why ten years, though, and not twelve, or five, or twenty-one? "Because we have ten fingers," he said, smirking. "Ten is a round number." Regardless, Rubin is likely to continue hunting for cosmic explosions and strange-looking stars long after its decadal mission has been completed—just as it will keep an eye out for any asteroids lingering in Earth's neighborhood. Compared to all its other ground-based telescopic family members, Rubin will be able to spot those pesky city killer asteroids anywhere in the sky, including at distances that other observatories would struggle to perceive. That attribute alone, one could argue, is worth the hundreds of millions of dollars it's taking to build and operate it.

Unlike the space-based NEO Surveyor, Rubin must deal with the atmosphere, which gets in the way of anyone hoping to do astronomy.

It also cannot position itself anywhere near as freely as an off-world spacecraft. And, being a multipurpose mission, it won't be spending all its time questing for undiscovered asteroids.

But the power of its snazzy camera means that it will be finding new NEOs left, right, and center during most of its runs. Ivezić told me that from the discovery of the first asteroid—Ceres, in the asteroid belt between Mars and Jupiter, spotted in 1801[30]—to the millionth discovery, it took 200 years. "When Rubin gets online, we will go from one million to six million in three years." And it's about to become fully operational. The pandemic delayed things, but with all going well, the big camera's "first light"—the moment it peers into space and sees glistening stars—should happen in the summer of 2024, with observations officially commencing in early 2025 at the latest.

Despite its remarkable capabilities, Rubin is occasionally framed as the supporting actor to NEO Surveyor's starring role. "There were a few recent meetings where NASA people were talking about NEO Surveyor, and they completely ignored Rubin Observatory, like we never existed," said a clearly peeved Ivezić. He had recently sparred with other astronomers who claimed that only NEO Surveyor could find asteroids interior to Earth's orbit, in the Sun's glare. Ivezić disagrees. "You can see Venus with your eyes. Obviously, you can see inside Earth's orbit."

But there is a reason why some astronomers see Rubin as secondary to NEO Surveyor in the asteroid-hunting sweepstakes. Like all ground-based observatories, Rubin is plagued by a problem that may soon become insurmountable: megaconstellations.

Samantha Lawler, an astronomer at the University of Regina in Saskatchewan, Canada, was pissed off. Preternaturally excitable, she spoke about her subject in allegro. All she wanted to do was look through a telescope at the solar system's multitude of weird worlds.

That is what sparks her joy. But by the time we spoke in 2021 and 2022, I could tell that an increasing amount of her firecracker energy had been aimed not at planets, but the artificial lights that had appeared in front of them.

Satellites have all sorts of purposes. We use them to navigate roads, cities, and countries. They can keep an eye on forests, oceans, skies, friends, and enemies. They help forecast the weather. They relay radio and cell phone signals. And they connect us to the internet. Earth's first artificial satellite, the Soviet Union's Sputnik 1, was launched on October 4, 1957.[31] Many more have joined it since. Some have malfunctioned, run out of thruster fuel, or were instructed by their human masters to die, tumbling back into the atmosphere and turning into metal confetti. But the total number of active satellites in orbit has continuously risen; in 2019, there were perhaps 2,300.

Just three years later, that number had jumped to 7,000. And just over half of those belong to the Starlink swarm. "That's insane," Lawler told me.

Starlink, a so-called megaconstellation of satellites, provides internet access to paying customers who install a small dish on their property. The aim[32] of the company that builds, owns, launches, and maintains these satellites—SpaceX—is to offer affordable, speedy internet access to those living far from a decent, cable-delivered internet infrastructure. Although it sometimes delivers on this promise, the fees the company demands from its customers suggest that the definition of affordability is open to interpretation. For example, as of July 2023, someone in New Mexico would pay a one-time $599 bill for the hardware, and then $120 per month for the standard connectivity option. For someone in poverty-stricken Haiti, it's $500 for the hardware and $50 per month for service.[33]

Many critics of the swarm go further: they have concluded that Starlink's ostensibly philanthropic goal—which has included providing internet access to nations impacted by disasters or war—is instead a cynical ploy* by SpaceX founder and CEO Elon Musk to earn as much money as possible, regardless of the consequences[34]—one of which is light pollution, and the potential downfall of much of Earth's ground-based astronomy.[35]

Satellites can reflect sunlight, sometimes adding little points of bright noise to telescopic images. Until 2019, these pinpricks had not intruded on astronomy in any major way. But that year, the first-generation Starlink satellites were launched. To provide high-speed internet connections to their customers, the satellites must orbit much closer to the planet than most other satellites. That makes the satellites prone to reflecting sunlight, turning them into white streaks graffitiing the night sky.[36]

If there were just a handful, even these bright streams would be a minor irritation. But from 2019 to 2022, Starlink satellites were launched in the hundreds. They are now the most common satellite type in orbit today, and they are defacing the darkness. One major issue is that billions of people with no say in the matter—including Indigenous peoples who rely on seeing the stars—are having their nightly canvas stolen from them without permission.[37] But Starlink has also proven disastrous for scientific astronomy. Every year since 2019, an increasing number of astronomers are examining their photographs and finding their observation runs tagged by Starlink streaks.[38]

SpaceX has attempted to dull their satellites. One version came with sunshades, which dimmed their brightness a little. But a subsequent

* Starlink satellites are manufactured at a facility in Washington State not far from Seattle. One day, Ivezić and a team paid them a visit. After meeting some of the staff, he asked one of them: Why are you so excited to be working on Starlink? According to Ivezić, they said: "No, we're all excited about other things. We are excited about working on a project that will send humans to Mars. And this is only a way to earn money so that we can support that other project." Ivezić told me, "This is the story that Elon Musk sold to his people. But there is nothing noble about Starlink."

generation of Starlinks were given lasers so they could directly communicate with one another in low-Earth orbit, an upgrade that necessitated the removal of those sunshades. The satellites of the following generation were again altered to reflect less sunlight, but nowhere near enough to stop leaving bright streaks in the sky.[39] Lawler noted that SpaceX employs some of the best engineers in the world. If the company wanted to make Starlinks fainter, they could do it. "It's just not a priority for them," she said.

SpaceX has also shown no willingness to slow down its Starlink proliferation. Unless their use is actively violent—which it is not—there is no international framework that puts enforceable boundaries on satellite deployments. Attempts to conjure up political or legal challenges to Starlink have, so far, failed. According to the UN, space is for everybody—which, in practice, means it belongs to billionaires who can afford to change its very nature.

As of July 2023, Starlink had 4,541 satellites in orbit.[40] The US Federal Communications Commission—the agency that permits or rejects the deployment and operation of US-based satellite swarms—has authorized SpaceX to operate 12,000 of them. Paperwork filed with an international telecommunications regulator suggests the company wishes to launch another 30,000.[41] Other companies vying for a slice of the same internet-providing pie, including Amazon's nascent Kuiper megaconstellation effort, hope to add their own thousands to the metallic crowds above our heads. In theory, as many as 400,000 of these satellites, owned by companies and nations across the world, could launch in just the next few years.[42]

Astronomers cannot keep pace. And megaconstellations could nix our chances to see a potentially hazardous asteroid barreling toward the planet.

Asteroid-hunting surveys like pointing their telescopes toward the Sun at twilight, hoping to spy asteroids interior to Earth's orbit that may be concealed by the Sun's glare. This is difficult enough as it is. For

NEO searches, "it's absolutely vital to get that twilight time," Lawler explained. Megaconstellations, chiefly Starlink, are literally blocking their view during that crucial time as they crest the horizon. "This is exactly the sort of survey that is going to be impacted," she said. "This keeps me awake at night. It's really quite worrying."

The Rubin Observatory, on paper, is an asteroid-discovering behemoth. But the reality is that Starlink and its cohort of low-hanging mirrors stands a good chance of clouding its sight. (SpaceX does not have a press office, and my attempts to contact them about this issue were met with silence.)

Future design tweaks may make Starlink satellites even more blinding than the current generation. "If they make them brighter than that, then the whole sky will look horrible," said Ivezić. "They would saturate our detectors." Based on their projected numbers, a catastrophe is plausible.[43] Even 50,000 somewhat dimmed megaconstellation satellites would mean that 10 percent of Rubin's nightly observations would feature a streak. For twilight surveys, almost every single photograph of the sky would include a bright, white line. NEOs will become harder to find, even for state-of-the-art Rubin.

"It's like the worst aspects of capitalism in every possible way," said Lawler. Starlink "is not a charity. It really does make us less safe as a species. This is the one natural disaster we could prevent if we detected it early enough, and we're screwing it up with our own technology. It's so frustrating!" Short of sending all NEO-hunting surveys into space—"I don't see Elon handing us $10 billion to do that," Lawler quipped—there appears to be nothing astronomers can do but put most of their planetary defense chips on NEO Surveyor.

Earth-impacting city killer asteroids are the unthinking villains of our story. Nobody wishes for them to triumph. That makes all seven billion of us the protagonists. But it has become clear that some human instincts—from our tendency toward being reactive instead of proactive, all the way to capricious self-interest—have slowed down our efforts to protect the planet. And no parable showcases the steepness of this uphill struggle more brutally than the death of Arecibo.

Even if you have never been curious about radar astronomy, you have probably seen an image of the Arecibo Observatory before—it was featured in the we-are-not-alone movie *Contact,* and made a lengthy appearance at the climax of the James Bond flick *GoldenEye,* when the characters, played by Pierce Brosnan and Sean Bean, battle while suspended over a gigantic dish carved out of the bedrock below. Those weren't fabricated sets or models; this facility exists—or to be more precise, it used to exist.

Arecibo has been Puerto Rico's scientific pride and joy since it was built into one of the island's natural sinkholes in 1963. Thanks to its record-breaking 1,000-foot dish, this colossal observatory was able to detect the faintest of radio waves being emitted by energetic sources many light-years away, from the auroras of exoplanets to the distant deaths of aging stars. But it was also equipped with a powerful radio transmitter, allowing it to ping planets, moons, and asteroids. Should an asteroid pass close enough to Earth above its cone of visibility, Arecibo could use its radar to precisely determine that asteroid's size and trajectory.

Making it through several earthquakes and hurricanes, Arecibo still had a lot left to offer to astronomy by the time 2020 rolled around. But on December 1, part of the instrumentation suspended above the enormous bowl failed and tumbled into the dish, leaving the observatory irreparably damaged. The cause of the collapse was clear: multiple cables had broken. But it's thought that the National Science Foundation, or NSF, which owned the observatory, was partly to blame; unwise architectural alterations and, in recent years, a laissez-faire attitude toward repairs and proper maintenance contributed to its demise.[44]

Astronomers were horrified. But for Puerto Ricans, such as Ed

Rivera-Valentín, a planetary radar astronomer whose career was both forged and fueled by the Arecibo Observatory, it was like witnessing the untimely death of an old friend.[45] Two years on, "it's still shit. It's a lot of shit," they said. "How does one deal with that?"

As was made clear by the 2022 North Carolinian asteroid impact war game, radar is essential for planetary defense. And Arecibo, to many, was the asteroid-pinging radar king. Consider the Goldstone Deep Space Communications Complex in California. The benefit of this radar facility is that its dish is mobile, so it can ping asteroids anywhere it can point. Arecibo's dish was fixed in place, so it could only see asteroids that passed directly above it. But Goldstone has a less powerful transmitter and a dish just 230 feet across, meaning it is unable to ping or see distant objects. That wasn't a problem for Arecibo, which could browse much of the inner solar system. "There's no radar telescope on Earth that will get you Mercury, Venus, Mars, anything else," according to Rivera-Valentín. "Goldstone will get you the Moon, that's it." Arecibo saw twice as many asteroids as Goldstone, making its collapse "a major loss for planetary defense."

The loss for Puerto Ricans was not just astronomical, but cultural. "When in Puerto Rico, and you think 'science,' you always thought of Arecibo. Having it disappear, the societal link to science was really abruptly cut," Rivera-Valentín told me. And, Rivera-Valentín added, it was no surprise. Puerto Rico is part of the US, but it isn't a state—it's a territory. That means that it doesn't have the political rights of states. A sole member of the House of Representatives speaks for the island's interests but lacks voting power, and the island's citizens—deemed American since 1917—cannot vote in America's general elections.[46] When troubled by disasters, from quakes to hurricanes, the island is not assisted by the federal government as quickly, efficiently, or intensively as it otherwise might be. Puerto Rico is perceived by some to be a de facto American colony,[47] and islanders debate which path, statehood or independence, they should try to walk down to change that ignominious status.[48] To them, Arecibo's collapse was another symptom of this political disease.

"You have all this baggage of colonial abuse. It's more of a vicious loss than an accidental loss, is the feeling most people have," said Rivera-Valentín. "People shouldn't have colonies, then abandon their colonies."

Although the December 2020 collapse was catastrophic, the NSF could have opted to rebuild. But in October 2022, the decision was made to not perform a resurrection—while also not proposing a similarly capable radar complex to fill the void Arecibo's death left behind. "Arecibo was absolutely still producing important data for asteroid science at the time of its collapse, and there is no replacement for its specific capabilities for high-resolution radar studies," Andy Rivkin, one of the DART mission's scientific leaders, told *Scientific American*.[49]

The NSF, a federal agency,[50] makes decisions based on the American astronomical community's collective feedback. It seems as if almost everyone wanted Arecibo to be repaired. So why wasn't it? In a word: bureaucracy.

Every ten years, American astronomers are asked what space-based query they would most like to answer, and how they would go about answering it, with a focus on where the community's funds should ideally go. The answers are collated into several major reports, including the Astronomy and Astrophysics Decadal Survey,[51] which focuses on things like stars and galaxies, and the Planetary Science and Astrobiology Decadal Survey,[52] which ruminates on worlds. The federal government looks at both and makes decisions while taking these recommendations into account, shaping the future of American astronomy far into the future.

These reports were the community's chance to officially declare their support for rebuilding Arecibo. The Astronomy Decadal, which is primarily supported by the NSF, asks scientists what recommendations they would make for America's ground-based astronomy facilities. But this report was being written up as the observatory collapsed, and technically the window for input from members of the community had already closed. That meant it was too late for anyone's views on resurrecting Arecibo to be included in the report.

That left the Planetary Decadal, which was still being deliberated on and composed. Rivera-Valentín, who was a contributor to that survey, explained that the community was able to highlight the loss of the facility and elucidate the hole its death opened in the fields of planetary science and defense. But they could not use this report to officially recommend that Arecibo should be resuscitated. The Planetary Decadal is primarily a NASA-focused feedback report. Per congressional edict, NASA must focus on space-based assets, and it is not allowed to consider or construct ground-based facilities. "That's the job of the NSF," said Rivera-Valentín. "So we couldn't say, recommend rebuilding [Arecibo], because NASA would say, 'Thank you but we're not allowed to, bye.'"

You would think that the explosive collapse of one of America's premier astronomic observatories would allow for some wiggle room. But no—rigid protocols prevailed. It was another example of how the country has not yet properly grasped planetary defense, which involves both space-based and ground-based assets that cannot neatly be segregated into two different boxes.

What happens now? "I don't know. I don't know!" Rivera-Valentín exclaimed. "There is no plan to fix the radar infrastructure."

The irony is that the latest Planetary Decadal, published in April 2022, is the first to feature a section about planetary defense. Chief among its recommendations is to get on with NEO Surveyor as quickly as possible.[53] But Arecibo, and ground-based radar, is noticeable in its absence. And we are all less safe as a consequence. "The lesson for me has been this," said Rivera-Valentín: "Bureaucracy plus science equals bad."

At its heart, that is what *Don't Look Up* is truly about: humanity is often its own worst enemy.

When the movie's comet is found to be heading to Earth, the Trump-esque White House reacts apathetically. They eventually get to work on a mission to use nuclear warheads to deflect the comet, but this is canceled not long after its launch when Peter Isherwell, the billionaire CEO of the BASH Cellular corporation, pitches the president on another idea: let BASH use their own spacecraft to disrupt the comet into harmlessly sized pieces. That way, when the fragments splash down in the ocean, they can be recovered, and their extremely valuable minerals can be mined for profit.

The revised mission fails, and the comet collides with Earth, killing almost everyone. It's a grim ending, for sure—but the takeaway message should not be that we are all doomed to die. The movie, Mainzer told me, was more optimistic than one may think. "It almost worked. The whole world came together and, in spite of all their differences, they almost mounted a successful deflection mission. And what happened was, one person, who had way too much money, and way too much influence, was able to get in there and mess it up at the last minute. And that's a fixable problem. That to me is the hopeful core of the movie. These are things we can choose to fix if we decide to."

At the movie's end, as the comet crashes into the Chilean coast, our main characters spend their last moments eating dinner together inside astronomer DiCaprio's house. With the house shaking, and with the lethal shockwave just moments from reaching them, he turns to his family, and says: "The thing of it is, is we, we really . . . we really did have everything, didn't we?"[54]

IX

The Cotton Candy Killers

September 2022 had arrived. In just twenty-six days, DART was going to impact Dimorphos. The star tracker on the spacecraft was still wobbling about, pushed by an unknown force, potentially jeopardizing the spacecraft's ability to navigate. The only thing that could be pushing something on DART was something else on DART, but trying to figure that out while the craft careened through space millions of miles from home was about as easy as juggling cups of boiling water.

Because Elena Adams and her team are the best at what they do, they figured it out. The problem began with—although wasn't the fault of—something called the power processing unit, or PPU, a type of box that provides electrical power to various or specific components on the spacecraft, including the NEXT-C thruster. Space is unfathomably cold, which meant that the electronics had to be kept relatively warm on the inside, lest they fail. That included the NEXT-C PPU.

Whenever the temperature of the PPU dropped too much, heating elements would kick into gear and warm it up. When the temperature was optimal—not too cold or too hot, but just right, like a bowl of perfect porridge—the heating elements turned off. That all worked as it was supposed to. But the physical panel to which the PPU was attached was good at conducting this heat—a little too good. That made it surprisingly reactive to the temperature changes, which caused it to physically change shape with each temperature fluctuation. And that caused the star tracker to be repeatedly nudged out of position.

Fortunately, this problem didn't require someone to catch up to DART and whack it with a wrench. A little bit of bespoke code, beamed to the spacecraft, managed to convince the PPU's heating element to switch on and off more frequently. This allowed the panel to remain at an essentially constant temperature, which nixed any major panel warping behavior. It may have taken almost six months to solve this problem, but they triumphed. The odds of DART hitting Dimorphos rose—but, still, no one could really say how close, or not, they were to 100 percent.

Naturally, I had to be there to see the drama unfold myself. Watching it on a laptop screen wasn't going to cut it. I wanted to be in *the* room, the one with all the desperately tense scientists presumably watching their spacecraft's final seconds—so I flew to Washington, DC, a short drive away from the Johns Hopkins University Applied Physics Laboratory in Laurel, Maryland. I had been to the capital once before and had thoroughly enjoyed drinking in all the sights. But this time, with all the eschatological reporting ringing in my ears, I couldn't help but think, *Wow, this city would be seriously easy to flatten with a Tunguska meteor.* Forget aggressive aliens with a predilection for destroying famous landmarks—a relatively small rock exploding in the sky would do it.

As it turned out, this was perhaps an unwise topic to bring up with my cab driver.

Nuclear bombs—sorry, NEDs—and kinetic impactors are not the only two ways to mitigate an asteroid strike. That instrumental SPACECAST 2020 report, the one in which Lindley Johnson first proposed that America should lead the way in the world's efforts to prevent such a tragedy, also featured a range of alternate planetary defense ideas.[1]

Some, at least at the time of writing, remain outlandish. One such example, sending swarms of nanomachines to masticate the asteroid until it is nothing but tiny bits, sounds more terrifying than helpful. Others, like antimatter bombs, are firmly within the realm of science fiction—for now, at least. The rest are as deceptively simple as they could be wildly effective, and some may be more elegant than a big explosion or ramming a spacecraft into the target. And perhaps one day, they will get tested out, too, following DART's imminent success—or failure.

Like paint. You could, well, just paint the asteroid.

It's nowhere near as bizarre as it sounds. Remember Sentry, and how it can work out how photons from sunlight, imparting the smallest of nudges to an asteroid, can change its trajectory? That's where the paint comes into play.

Nobody is quite sure how it would be accomplished, but it would likely involve an interplanetary Banksy—a robotic spacecraft that would rendezvous with the asteroid, OSIRIS-REx style, then spray-paint one side of it. It would need to be a special sort of paint; normal paint would be vaporized in the zero-pressure, ultracold environment of space.[2] Paint one side white, and photons from the Sun would bounce off it at a much higher rate, giving it a very gradual push. After a few decades, you will have changed that asteroid's orbit enough to potentially push it out the way of Earth.[3]

Asteroids often misbehave. Some rotate slowly, but others spin like hyper-caffeinated ballet dancers. Working out how to use reflective paint to move an asteroid effectively and safely out of harm's way may not always be possible—and even if you could make the numbers work, you would need a very long warning time for this method to be considered.

On the other end of the technological scale, we have ion beams. Ion beams may sound like something from *Star Wars* (and they are), but they already exist. DART's NEXT-C thruster is an ion thruster, and similar versions of it have appeared on countless other spacecraft and satellites. Most use xenon, a nonreactive gas, as fuel. Electrons, the particles that orbit atomic nuclei, are fired at the xenon, knocking off the xenon's electrons and producing positively charged ions. These ions are then attracted to a negatively charged component and are angrily ejected from the thruster at 90,000 miles per hour, creating a beam of ions that gives the spacecraft thrust.[4]

So, some have thought, why not use that thrust to push an asteroid out of the way? It's a more delicate manner of deflection than a kinetic impactor, and if you can park a spacecraft next to the asteroid, you can adjust the ion beam's deflective power and angle over time. Sounds great, but this would require some extremely precise spaceflight over a long period of time, perhaps several decades' worth.

During one of our conversations, Olivier Hainaut, our astronomer friend from the European Southern Observatory, asked: "Have you heard of the gravitational tractor?" This was a surreal question to be able to say "yes" to. It sounds way more sci-fi than it actually is; there is no Death Star–like tractor beam at play here. This method would involve parking a heavy spacecraft near the asteroid and using the gravity of said vehicle to tow the rock very slowly out of the way. Again, that would take some deft autonomous piloting by the spacecraft. "The technology works on paper," Hainaut explained. "How you park near an asteroid and give just the right amount of impulse . . . ehhh." He shrugged. "It works on paper. It would be good to experiment with it."

If you'd like your method of planetary defense to be a little more offensive, then why not try *The Expanse*? This tactic was deployed in an excellent series of novels, and a subsequent TV series, where various human factions are engaged in acts of subterfuge across Earth, Mars, and the asteroid belt. Scientists love it, in part because it treats space as we know it, with all the appropriate laws of physics intact. As an example, those living on low-gravity asteroids cannot simply move to Earth because the significantly stronger gravity crushes their bones.

Spoiler alert: at one point, extremists from the asteroid belt paint several city killer asteroids in a special stealth coating, then fling them onto Earthbound trajectories—a futuristic version of the September 11 attacks, where one asteroid impact is thought to be an accident until the second one hits.* The killer asteroids in *The Expanse* are propelled toward Earth using tech similar to gravity tractors, raising the concern that you could use them to tow an asteroid onto a dangerous path to attack a hostile country. In the show, however, Earth does eventually find a way to spot these rogue asteroids, and ultimately blasts them into tiny pieces with a mixture of ground-based railguns—a weapon that essentially uses magnets to propel a projectile at extremely high speeds—and space-based torpedoes. Amazingly, railguns already exist, although the US Navy is having difficulty implementing the technology.[5] But as a way of defending the world against asteroids that we don't see coming until the last moment, it isn't difficult to imagine railguns going the way of nukes: starting as weapons of war, then being coopted for planetary defense.

Perhaps we're thinking about this the wrong way around. Instead of firing hypervelocity projectiles at the asteroid, we could always turn an asteroid's remarkable speed against it. That is at the heart of a relatively new idea, PI Terminal Defense. PI stands for "Pulverize It!" And the best part is that we would get to ambush the asteroid.

PI Terminal Defense is the brainchild of Philip Lubin, the director

---

* In 2021, for the *New York Times*, I covered a study that involved throwing simulated asteroids at Earth to see if last-minute NEDs could properly disrupt them. My editor, the marvelous Michael Roston, loves *The Expanse* as much as I do—so we compared the lead author of the study to the TV show's asteroid-throwing antagonist, Marco Inaros. The actor who played said villain enthusiastically approved of the article, while speaking in-character, on Twitter. My geek cred skyrocketed.

of the Experimental Cosmology Laboratory at the University of California, Santa Barbara. He has a background in theoretical physics—not the field usually associated with planetary defense. No matter, he told me. "Almost everything in life interests me. Except politics."

Perhaps the conception of this mitigation method was inevitable. Lubin spent many years at the Lawrence Berkeley National Laboratory in California, a hub of inventive physicists that has collected a staggering sixteen Nobel Prizes to date.[6] It was also once headed by Luis Walter Alvarez, the experimental physicist who, along with his son, Walter, proposed the hypothesis that an asteroid wiped out the dinosaurs. "It was a very free-flowing intellectual environment," said Lubin. Being there may have sown the seeds of PI in his mind, because years later, in 2015, he began to wonder if he could "use Earth's atmosphere as a bullet-proof shield." His idea was developed further in 2020, when the onset of the pandemic gave him extra time to think.

So: What is PI? As the name suggests, it is a way to disrupt an asteroid into millions of smaller pieces. Rockets would be launched into space and head toward the asteroid—but each one would stop somewhere along its path. Then, its payload would be unleashed: a series of metallic rods. They would float into space, forming a lethal roadblock. And there they would remain until the asteroid, moving at tens of thousands of miles per hour, runs into them, shredding itself into pieces no larger than thirty feet. As the phrase "Terminal Defense" implies, this method could be used as an emergency measure: according to analysis by Rubin and his colleagues,[7] a Chelyabinsk-size asteroid could, if spotted in time, be disrupted just one hundred seconds prior to impact, with the smaller chunks burning up harmlessly in the atmosphere. A Tunguska-size asteroid, they reckon, would require disruption five hours prior to impact.

But this need not be thought of only as a Hail Mary measure. "I'm not an advocate of playing chicken," he said—you could deploy these rods at any distance from Earth, perhaps decades before an Earthbound asteroid could manage to reach us. And, at least on paper, PI would work to disrupt larger asteroids, the sort that may take multiple kinetic impactors to deflect, or those that simply may be too massive to knock out of Earth's way in time. "For large things, [deflection's] just not viable," he said.

Perhaps Rubin's idea will get NASA funding someday. Hopefully so, as we need as many effective tools in the toolbox as possible. But there is a reason that the kinetic impactor technique is being tested first and is so widely supported in the scientific community. Mounting a kinetic impactor deflection mission is technologically and politically more straightforward than trying to design a deep-space, NED-driven deflection or disruption campaign.

Kinetic impactors just make the most sense most of the time, and it doesn't take a preeminent planetary defense researcher to come to that conclusion. In the 1950s, a mile-wide asteroid named Icarus was forecasted to make a close flyby of Earth in the summer of 1968. One year prior to the flyby, Paul Sandorff, a lecturer in space systems engineering at the Massachusetts Institute of Technology, thought it would be a fun exercise to give his students a very different sort of assignment: How would you prevent Icarus from destroying the world?[8]

They were given four months to solve the problem and told that money would be no object. They were handed information from astronomers (including the discovery of Icarus) and data from nuclear weapons experts, and some students were flown out to an air force base to check out some potentially useful rockets.[9]

After months of hard work, the students presented their conclusions to Sandorff, various aerospace engineers, and a journalist from the *MIT Technology Review*. They proposed a two-pronged mission: four of NASA's giant Saturn V rockets would slam into Icarus to deflect it, with two more—this time, armed with NEDs—following later, to make sure the deflection was successful. There would be some disruption, but they had proposed a chiefly deflection-focused venture.[10]

Incidentally, the assignment resonated with those outside MIT: word spread, and it received national press coverage. The story was eventually picked up by Hollywood, who used it as the basis for their 1979 sci-fi movie *Meteor*, in which Sean Connery stars as a scientist hoping to stop an asteroid from impacting Earth. Closing the loop, MIT's *Tech* publication critiqued the film. "An intellectually astute audience such as the MIT community will find a number of flaws in *Meteor*," the reviewer wrote,[11] including "an American scientist who has the accent of a native Scotsman. Nevertheless, the movie is a good disaster thriller."

In the year leading up to DART, everyone I spoke with had plenty of confidence in the effectiveness of the technique, regardless of whether DART itself would hit Dimorphos. Partly, this approval was because of the simplicity of this form of planetary defense: if you hit something in space, even an asteroid that turns out to be more like a collection of cozy boulders than a single rock, it should react by flying in another direction. The conviction stemmed from the many experiments on Earth that previewed the technique. Many of those trials involved virtually recreating the impact. Some of them involved pointing a big gun at the shards of asteroids and shooting them.

Every now and then, you stumble across something, or you make something work, or you say or do something, and you experience a very specific type of elation: what just happened was incredible, and you thank yourself for all the choices that led to that moment. Peter Schultz has probably had scores of these moments. But I suspect that one of them was when he used a space-age cannon to fire a bullet at some cotton candy—you know, for science. "It went right through the cotton candy. And because cotton candy is also very explosive, the whole thing exploded," he told me, as if it were the most normal thing in the world. Out of all the firing runs he had done over his multi-decadal career, “that’s probably the oddest one,” he said. It was for *National Geographic*. “Every time I did a TV program, I would do something stupid and ridiculous.”

If planetary defense has a resident *Back to the Future* Doc Brown type, Schultz would be a top contender. He’s a planetary geologist and impacts expert at Brown University, and the overseer of NASA’s Vertical Gun Range, or VGR, at the Ames Research Center. His inventiveness and zaniness do not come across in the way he speaks; he is calm and measured for the most part. But the image of cotton candy murder speaks considerably louder than words.

Ames was founded nearly a century ago and is home to an array of bleeding edge experimental laboratories,[12] from giant wind tunnels to spaceflight simulators, biotechnology research hubs, and nanotechnology facilities. It is also home to supercomputer-wielding scientists that work out the best ways for spacecraft to safely reenter the atmosphere—and the worst ways in which an asteroid’s airburst could maim millions.

It is also where one can find that aforementioned vertical gun. A strange, orange, metallic giraffe stretches from the building’s ceiling to floor. In between are a series of tubes, switches, knobs, and pipes, combining to form a crown atop a pale blue chamber emblazoned with the famous NASA logo. That giraffe structure holds on to a fourteen-foot barrel, the gun-like apparatus that propels what are effectively bullets toward unsuspecting targets in that blue lower chamber. But the VGR looks nothing like what anyone would imagine a gun to resemble.[13] It was pieced together a lifetime ago during the early days of the Cold War from all sorts of bits and pieces. Part of it once held rockets designed to shoot down Soviet nuclear missiles.[14] Today, the curious-looking contraption looks like it would make a very interesting casserole.

But it is, in fact, a gun. Gunpowder charges are used not to launch bullets, but to violently squash a pocket of hydrogen gas contained within that enormous barrel. This high-pressure bubble has no choice but to escape its confines, and so it whooshes down the tube. Inserted at the business end of the barrel is the scientists' projectile of choice—metal, mineral, ceramic, or glass-based, of any shape or size. The gas rushes to meet the projectile, blasting it out at thousands of miles per hour and sending it flying into whatever target is placed inside that lowermost chamber[15]—all accompanied with a satisfying thud, bang, or boom, depending on the nature of the impact.

Despite its complex-looking design, using it appeared to be relatively simple. Prime the gunpowder charge and the gas—check. Put in a bullet. Place a target inside the impact chamber. Seal everything up. Change the position of the giraffe and the affixed barrel to alter the angle that the bullet will hit its target.[16] Check the cameras are working. Stand somewhere that isn't right next to it. Press a button. A resounding, tinny thud rings out. Done. Very nice. And, unlike the explosive experiments I took part in as part of my PhD—involving the burial and detonation of TNT in the middle of a field in Buffalo, New York—there was very little post-blast cleanup required. I could get used to this, I thought.

Bizarrely, the original purpose of this gun was related to the Apollo landings. Not because anyone was concerned that Moon aliens would try to kill astronauts Neil Armstrong and Buzz Aldrin, and they needed a way to fight back; scientists were unsure how the lunar soil would respond to being walked over, with some worrying that it may swallow the astronauts or their landing module up like quicksand. So the VGR was pieced together in 1966, and scientists fired at simulacrums of Moon soil to estimate the depressiveness of the lunar surface.[17]

The device was going to be mothballed after that, but other scientists protested, pointing out that the VGR could help them understand asteroid and comet impacts across the solar system. They won out, and in 1980, Schultz began running his own experiments on it. More than forty years later, the VGR is still running strong. Other hypervelocity gas-propelled gun designs have popped up elsewhere, from the UK to Japan. But the VGR is still in vogue to run impact experiments. And Schultz is still not remotely tired of his work.

"Could you believe that anything could survive a high-speed impact?" he asked me excitedly. After all my reporting, I could only answer in the negative. But he told me about a nine-million-year-old impact site in Argentina that was recently examined, one momentous enough to melt the ground and turn grains into glass. Improbably, plant material was found imprisoned within that glass, "preserved perfectly," said Schultz—in glass that cooked at over 3,000 degrees Fahrenheit. "An impact should destroy everything." The question, then, was obvious: What the hell had happened?

Curious, Schultz deployed the VGR. He took some pampas grass, a common plant in Argentina, and put several blades on a porous pumice target. He loaded the gun with bullets of glass and fired them. Spooling through the footage recorded by high-speed cameras, he discovered the answer to the Argentinian riddle. The glass bullets impacted at such extremely fast speeds that both the bullets and the target turned into liquid glass, replicating what had happened on that destructive day long ago. The impacts that happened at oblique angles annihilated some of the target, but other parts were spared the highest temperatures and pressures. These less scarred sections were soaked in turbulent waves of molten glass, which enveloped, rather than ruined, the target matter, trapping pieces of grass inside. "I've actually fired bacteria into this stuff, using a hollow sphere," Schultz told me. "It's a paper I want to write. It's called 'Shooting the Shit.'"

Experiments at the VGR, whether conducted by Schultz or other planetary scientists, have offered clues to the aggressive sculpting of the solar system, including Earth. They have revealed that impacts are messy—really messy. Schultz brings up the exciting-sounding notion of meteor decapitation. If an asteroid makes it to the ground, much of it may bury itself. But the back end of it may snap off and keep going, ultimately impacting many miles farther downrange. That has important, and slightly disturbing, implications for planetary defense.

Possible crater features, those that may be used to identify whether a comet or an asteroid was its excavator, have been another of Schultz's obsessions. Surprisingly, there is no clear way to tell if an impact crater was made by a comet or an asteroid—unless you find a big chunk of the surviving impactor, which is rare. And not knowing is a problem. "The assumption has been that comets represent less than 4 percent of the impacts that occur on Earth," Schultz said. That assumption may be wrong. That asteroid impacts are more common is incontestable, but are cometary impacts quite as uncommon as we think?

Some headway has been made in recent years: field expeditions to craters, astrophotography of craters on other worlds, computer simulations, and experiments at the VGR have been combined to suggest candidate diagnostic features. Perhaps, as one study suggests, certain streaks of matter stretching out of craters may be created only by comets.[18] But for now, nobody can say with any great specificity how frequent asteroid and comet impacts are. Not knowing quite how likely we are in any given century to be hit by a Tunguska-size object is a little unnerving—and makes me wonder, yet again, why missions like DART and NEO Surveyor are only just being realized now. On the bright side, at least they *are* happening.

Experimental runs by others have also proffered pointers to future kinetic impactor missions. Previous tests of this technique, including many virtual recreations of various deflection scenarios, assumed that the asteroid would be made of stuff similar or identical to a volcanic rock named basalt. It's an extremely common rock type on Earth as well as in space, so it was a reasonable assumption. But as space missions from Hayabusa2 to OSIRIS-REx have demonstrated, asteroids are often weakly bound entities, not monolithic monsters.

Hoping to find out how different asteroid types would respond to kinetic impactors, George Flynn, a physicist at the State University of New York in Plattsburgh, was left with no choice: he had to shoot not at recreations of asteroids, but genuine pieces of asteroids—meteorites, in other words.[19] "If you want to deflect an asteroid, you really don't want to use data on terrestrial rocks," he told me. So he bought some, which wasn't exactly easy. You cannot just phone a university or museum and ask to put their extremely valuable meteorites in a cannon designed to shatter them. But his proposals proved convincing enough to certain institutions, some of which sold him meteorite pieces at $1 per gram—a little over three-hundredths of an ounce. "In order to do the recoil measurements, we have to shoot samples that are close to one hundred grams," Flynn explained—a small price to pay for bolstering the planetary defense cause.

After firing bullets at various meteorites in conditions mimicking deep space—a vacuum, with impact speeds matching expected spacecraft versus asteroid levels—they found that certain asteroids are more prone to fragmentation than others, especially carbon-rich, porous, and soggy ones, suggesting accidental disruption during a deflection campaign is a significant possibility. And common asteroid types are twice as tough to shatter as terrestrial volcanic basalt,[20] which could be an issue for a real-life disruption mission.

Schultz and other scientists will continue chipping away at problems like these for as long as the VGR still has life in it. Yes, he will spend time on the more serious matters relating to planetary defense and the past, present, and futures of worlds. But that won't stop him exploding saccharine treats every once in a while. "This is just so much frickin' fun," he told me. "If I have a lot of people think I'm crazy, let them think I'm crazy. It buys me more time."

Schultz's storied tenure as the VGR's science coordinator undoubtedly opened new avenues of planetary science, including planetary defense, that would have otherwise eluded scientists and engineers, forever changing the field for the better. But his greatest contribution may be his ability to foster or inspire a countless number of students—some of whom found their way to working on the DART mission.

Unsurprisingly, the VGR was a vital source of information for the Deep Impact mission in 2005. Hoping to get a preview of what effect the spacecraft's copper bullet would have on Comet Tempel 1, scientists descended upon Ames Research Center, set up some experiments, and blasted holes in cometary clones. "I made a suite of experiments," said Schultz. "Look, if you see this, you'll get that." And in the end, the VGR's visuals were used to explain the unexpectedly spectacular results of Deep Impact's one-in-a-million shot.

Jessica Sunshine, an asteroids, comets, and impacts expert on the DART team, was a colleague of his at the time. "Jessica was standing next to me when Deep Impact hit. And she turned around and picked me up about three feet off the ground," he said.

Megan Bruck Syal soon became another acolyte. "The whole reason I'm in this field is because Pete Schultz gave a talk," our resident NED expert told me—a lecture at her undergraduate college, showing off the results of the Deep Impact mission. Given that he's such a prominent name in the community, as well as an impacts expert, I was curious about a decision Schultz made regarding DART: he would not be there in person at the Applied Physics Laboratory on September 26, the day DART was destined to die. Why? "I had six of my students on that team," he said. "And I decided it was their time."

Angela Stickle, the impact simulation queen of the DART mission, was also one of his students. Schultz may not be physically present, but his legacy would be. "There's a thing we do at NASA Ames whenever you shoot the gun: cross your fingers, cross your arms, and cross your legs," she told me just days before I showed up in Maryland. "If you don't, and something goes wrong, then it's your fault."

This minor act of contortion is called the Gault position, named after another of the VGR's pioneers[21]—and it's a habit still practiced by Schultz. DART may have been powered by precision engineering and rock-solid science, but everyone agreed that a little bit of luck wouldn't hurt.

On September 11, 2022, a day before I arrived at the Johns Hopkins University Applied Physics Laboratory, a little hatch popped open on DART's side, and a spring-loaded mechanism sent a toaster-size stowaway into deep space. LICIACube, the Italian microsatellite that would bear witness to the explosive triumph—or withering failure—of DART, was set free.[22]

Like every major milestone in the prototypical planetary defender's journey, it was a little tense for Elena Adams and her crew. "You'd opened this giant maw into space," she told me. Lots of heat rushed out. The cold crept in. Nobody wanted any more temperature fluctuations or instrument malfunctions, and, to their relief, none were detected—even after quadruple-checking the dashboard stats just in case an error was buried among the cacophony of positive-sounding bleeps and bloops.

DART was fine. But LICIACube decided to make its debut with unnecessarily dramatic flair: it should have electronically waved hello to Earth the moment it was liberated, but no such signal was received. Minutes went by. Nothing. "We were all on pins and needles," said Adams. Then, after forty-five minutes, there it was—LICIACube was healthy and reporting for duty. Goddamn you, you adorable metal box.

With so many plot twists, I wondered if Adams had had even a single moment where she wasn't going dizzy from trying to keep track of a thousand different engineering conundrums all at once. Not long before I arrived in Maryland, she told me that it depends. "When you're sitting with friends with a glass of wine, you can wax philosophical as much as you like," she said about the long-term, world-saving goals of DART and its technological descendants. But day-to-day, right now, there was one overarching query that ruled her existence: "Is this actually going to work?"

On September 12, on a characteristically muggy day, I was in a conference room inside the Applied Physics Laboratory, along with a few dozen other journalists, listening to several members of the DART team, who sat on a curved table before us. This was the beginning of the festivities for the press. Impact day was fourteen days away. NASA's Thomas Zurbuchen, the associate administrator for the Science Mission Directorate, stood before us. The very next day, NASA officially announced that he would be leaving the space agency for shores unknown.[23] Zurbuchen's tenure included the actualization of many paradigm-shifting missions, including DART and, eventually, NEO Surveyor—funding kerfuffles aside. Dr. Z, as he is affectionately known, had garnered the respect of much of the community[24]—and at the dawn of DART's history-making month, he was an appreciated presence, someone clearly enjoying the electric atmosphere. DART was part of his swan song.

Zurbuchen knew that September 12 held special significance for those keen to venture to the stars. "At 10 a.m. local time in Texas, in Rice Stadium, in front of 40,000 people, President Kennedy gave a speech," he said. The part many will recall is this: "We choose to go to the Moon in this decade and do the other things, not because they are easy, but because they are hard." It was the declamation that culminated in America beating the Soviet Union to the lunar surface.[25] But that wasn't the section of the speech that Zurbuchen was interested in. He recited: "We set sail on this new sea because there is new knowledge to be gained, and new rights to be won, and they must be won and used for the progress of all people." He looked up. "And that's the theme for today: for the progress of all people."

For the first time ever, humanity was going to try to do something previously ascribed to the hands of unseen deities. They were going to shift the stars, just a little, to prove that the world can be protected from an unthinking cosmic force. It was exhilarating to be there, but as we were shepherded out of the room and toward mission control, Adams's eternal question began echoing in my mind. Is this actually going to work?

X

Ashes to Ashes

Inside DART's mission control room, Elena Adams was holding court, standing in front of a handful of reporters, and flanked at some distance by a series of stuffed animals. I didn't know the story behind these furry friends. But it instinctively made sense to me, mixing the grandiose with the silly—even when you're talking about a mission that's going to practice deflecting the sort of asteroid that could destroy a city.

Someone asked Adams how hard it would be to hit Dimorphos. Deep Impact hitting Comet Tempel 1 had been far from easy, but at least the target was almost four miles long. Dimorphos is just 520 feet across, moving through space at tens of thousands of miles per hour. "Imagine you're at the JFK airport, and you know where the Dallas airport is. And you have a dart. The dart itself is only two-and-a-half millimeters at the tip; it's tiny. And you take that dart from JFK, and you throw it to Dallas, and you hit the center of a bullseye on the actual dartboard. Except you don't know where the dartboard is within the Dallas airport," Adams explained. "So that's how hard it is to do it."

Mission control rooms in the real world look like you'd imagine:

there are consoles and seats, and people staring at screens. It's not too different from how they're depicted in movies. But the monitors on the walls immediately reminded me of the dashboard of an X-Wing fighter—you know, the workhorse of the Rebel Alliance, the one that dared to do the trench run and destroy the Death Star? If you don't know what I'm talking about, then firstly—how have you not seen *Star Wars*? Secondly, imagine the sorts of graphics you would have seen in an arcade game in the 1980s, with lines and circles and dots representing space stuff. That's what those screens looked like, some of which were tracking DART's movement through space toward a target.

Despite some anxiety about whether the impact would happen, Adams was having plenty of fun previewing it. Dimorphos, lest we forget, was a largely unknown entity; apart from a vague size estimate, nobody knew much else about it. Someone asked Adams what she thought it would look like, from DART's unique perspective. "It'll get bigger and bigger, and then, boom," she said. Fair enough.

I knew that various team members had made bets among themselves about how much Dimorphos would be deflected. The minimum threshold was to reduce the asteroid's 11 hours 55 minutes–long orbit around Didymos by just 73 seconds,[1] but some scientists were optimistically hoping for an adjustment closer to 10 minutes. A greater time change would imply that most asteroids of this size could be knocked back by a relatively small mass, something that would bode well for defending the planet—especially if we have little warning time. But I was curious: Had bets been made about the appearance of Dimorphos?

"We did talk about starting a pool," said Adams. They hoped it wasn't something too weird, like a dog bone shape; that would make aiming at the center more troublesome for SMART Nav. "These are kind of the pathological shapes . . . that and the donut, which we've trained ourselves against." Ah yes, the donut: a ringed asteroid with a hole in the middle. Improbable, perhaps—but not impossible, for the universe is a demented sculptor.

Adams explained that they simulated a donut scenario with DART to test its mettle. Just because they could, they transformed Didymos into the Death Star, with the Dimorphos donut emerging from behind it. Once again, *Star Wars* had found its way into the planetary defense saga. Adams thanked Evan Smith, one of her engineers, who came up with the idea. "We flew right through the middle of the donut," she said. "We hit, but we didn't." A donut-shaped Dimorphos would be a bit of an issue, it seemed.

Suddenly, one of the fluffy creatures became thrillingly relevant. Adams explained that DART has its own mascot. "And it is DART Vader." She held up one of the plushies: it looked like a very adorable, pint-size Darth Vader, with two white eyes of slightly different shapes gleaming on his helmet-covered face. The asymmetric eyes represented Didymos and Dimorphos. Smith "was dressed as DART Vader too, for some of this," said Adams. That may have been the greatest thing I had ever heard. Who says you can't have fun while trying to save the world?*

What about other scenarios in which the mission does not go according to plan? They had practiced nearly two dozen of them, Adams explained. What would happen if the communications relay between Earth and the spacecraft was severed at an especially awkward moment? What if the asteroid was so dimly lit that DART had difficulty tracking it? These were relatively easy hurdles to leap over.

---

* So, what was with all the other mascots? They were all either good luck charms brought in by various team members or, in some cases, the floofy faces of past missions. "The fuzzies on the left are llamas from the solar probe team, from the Parker Solar Probe," said Adams. Why did they have a llama as a mascot? "I don't remember."

Others were more problematic, including the possibility that DART would maintain its initial lock on the larger Didymos, and refuse to switch its flightpath toward Dimorphos.

What if they missed? For a moment, Adams lost some of her seemingly irrepressible zeal. "We need to make sure the spacecraft is safe," she said. They would switch DART out of its kamikaze run, make sure it was saving what it had left of its propellant, and then hopefully command it to swing around the inner solar system in preparation for another impact attempt. "It will take us another two years to get back and try [to hit] Dimorphos again." October 31, 2024: "Because we like to hit all the major holidays."

An impact in 2024 would still be a success—but a lesser kind of victory, and clearly a letdown for everyone hoping to make a needle-shifting splash on the first try. Just a few weeks before my tour of mission control, this contingency seemed highly unlikely; the whip-smart algorithmic brain and nimbleness of the spacecraft meant that September 26, 2022, appeared destined to be DART's death day. But an impact felt all-but impossible when, on September 7, NASA announced that DART had gotten its first good look at Didymos[2]—and its target asteroids looked so ridiculously tiny they may as well have been invisible. By this point, the spacecraft had traveled many millions of miles on a circuitous, circumsolar racetrack toward its goal. A composite shot made of 243 separate images, taken by DART's eye, DRACO, depicted Didymos as nothing more than an extremely faint dot 20 million miles away, a snowflake lost in a diamond-suffused ocean. That the binary asteroid could be seen at all was good news: DRACO was operating as it should be. But that shot reinforced Adams's analogy—hitting this small asteroid was like throwing a dart across America to impact a minute target that nobody could see.

At the press conference just before the mission control tour, I clocked Smith, the *Star Wars*–loving engineer—and he looked almost distressed, which was perfectly understandable considering how close we now were to impact day. Adams was a little nervous, too, but she held her extremely talented team of engineers together so well it looked deceptively breezy. Having fun with the mission certainly played a role in quashing some of the anxiety, and she wasn't the only one to be in that mindset. On September 6, Andy Rivkin—one of the lead scientific investigators on the mission team—sent out a tweet[3]: "Ahoy, mateys! The #DARTMission is under 500 hours from its arrival at Dimorphos. If you were onboard the #DARTMission right now, you'd see the Earth as by far the brightest (non-Sun) thing in the sky, about the size of the letters down at the bottom of a typical eye chart, but only about 25% lit." He added, "I'd also suggest you get off DART ASAP. ;)"

September 22. Four days until impact.

The last two weeks had probably been occasionally fun for the DART team. But by September 22, many of them just wanted it to be September 26. Enough of the drama—let's get this done. That was the collective attitude of the engineering side of the mission: this is awesome, but we want to succeed, celebrate, then sleep for a long time. The scientific observers, unable to control anything and clearly impatient in their desire to see that impact happen, generally sounded more excited than nervous.

That day, I sat among several reporters in an auditorium at NASA headquarters, home to the Planetary Defense Coordination Office; we had been promised an update on DART's progress. Among the speakers were the planetary defense officer Lindley Johnson, and, of course, Elena Adams. Once again, they wanted to emphasize just how difficult this mission was—not because they were especially concerned that they would miss, but because it remained a possibility. Space, as the oft-repeated adage goes, is hard. So is creating a planetary defense system.

Adams explained that, in the coming hours, a few manual trajectory correction maneuvers were going to be sent from Earth to DART. She underscored that what everyone would be seeing from Earth would be almost real time, but not quite: there would be a 38-second communication delay, so if DART hit its target, we wouldn't know until 38 seconds after it happened. That delay also meant that an hour or so before impact, the team would have no choice but to take their hands off the proverbial wheel and trust that SMART Nav would be able to fly to Dimorphos all by itself.

Making manual corrections to the flight of a spacecraft so close to impact day proved a little controversial, Adams told me after the press conference. Some of her engineers expressed caution. "It's usually unheard of, you would not do that," she was told. But she felt that it was worth the risk of a thruster malfunction or a navigation error if it put DART on as precise a terminal flight as possible. "No, we're gonna do it, because they'll put us on the right trajectory," she insisted to her team—and they relented. DART shimmied about millions of miles from home, just days before it was due to make its permanent mark on the solar system.

When Johnson spoke, he didn't mess about. "This is a very exciting time, not only for the agency, but for space history, and for the history of humankind, quite frankly," he said. "The first time that we are able to demonstrate that we not only have knowledge of the hazard posed by these asteroids and comets that are left over from the formation of the solar system, but also have the technology that we can deflect one on a course to impact the Earth. This demonstration is extremely important to our future here on the Earth, and life on Earth."

On September 25, DART was 330,000 miles away from Dimorphos, and closing in at 14,000 miles per hour. Even at such a relatively close distance, Didymos, the far larger asteroid of the binary system, would have been twenty times too dim to perceive with the naked eye from the spacecraft.[4] Swimming in a soup of delirium and apprehension, many of the team were unable to sleep.[5]

That evening, I sent messages to both Adams and Rivkin, hoping to get a read on how everyone was doing—those whose jobs would end upon impact, and those whose work would become extremely intensive. Rivkin was chipper. "Everyone is raring to go here. We've been having daily Investigation Team telecons, we've got people at observatories around the world, everything looks good with the spacecraft," he wrote me.

"We are feeling ready," Adams said, via email. "And ready to get it over with! :)"

September 26. At about 7 p.m., DART was due to die. This was it: it was impact day—or the oh shit, we missed day. In the past couple of weeks of reporting I had met another journalist, Zack Savitsky, who was covering the DART mission for *Science*, and we decided to travel to the Applied Physics Laboratory together for impact day. When we arrived just after midday, journalists were everywhere, some furiously writing notes, others talking directly to camera in a number of languages. Screens positioned up high featured NASA spokespeople pointing at DART paraphernalia, explaining the science of the mission, all while a countdown ominously ticked away in the corner of the screen.

Reporters are supposed to maintain an objective distance from the story they are writing, but that was simply impossible when it came to DART. This was the culmination of decades of work, and its success would make the entire world safer. A miss would make for a melancholy but narratively interesting denouement to the tale. But if any of my stories were to have a happy ending, this was by far the most deserving. If DART wins, everybody wins.

Several small press briefings kicked off the afternoon. NASA paired up various members of the DART team and sat them on elevated stools in front of small lakes of seated journalists, and it initially looked like the speakers were about to engage in slam poetry or improvisational comedy—which, of course, would have been totally acceptable. Sadly, but I suppose more appropriately, everyone leaned into the profundity of the situation. "We're going to do something today that's never been done before. We're going to alter the course of a celestial object," said Bobby Braun, the head of the Applied Physics Laboratory's Space Exploration Sector. "I thought about that on my drive in, and it gave me goosebumps." This was, he said, something that "has importance to the entire Earth." He wasn't wrong.

At 3:08 p.m., Braun explained that the spacecraft was just a few minutes away from becoming fully autonomous—the moment when DART would become independent and fly itself to its first and final destination, though emergency human inputs from the ground would still be possible for a little while longer. The celestial cartographers at the Jet Propulsion Laboratory were just doing their final checks to make sure DART was heading toward Dimorphos.

Andy Cheng, the originator of the DART mission concept, was also present—and he looked skittish. He told those in attendance that he was anxious, constantly thinking about something he may have done wrong during his work on the planetary defense test. Like what? "Anything," he said. At least this mission, unlike most others, had had a short conception-to-completion time—a mere eleven years, rather than multiple decades. Overall, it had been "incredible fun."

During her session, Angela Stickle, DART's impact simulation maestro, revealed something wonderful. Per international rules, astronomers cannot name newly discovered features on other worlds whatever they want. There is no valley on Mars named Dave, no cloud formation on Saturn called Priyanka. Each major celestial object has a theme.[6] Many of the moons of Uranus are named after Shakespearean characters,[7] like Oberon and Miranda. The monikers of volcanic features on the Jovian moon Io are taken from the fiery gods of various religions and from places and people in Dante's *Inferno*. Pluto and its moons take inspiration from mythological underworlds and their characters.

It was agreed that the features on Dimorphos would be named after percussive instruments, such as Dhol Saxum, after the Latin word for rock (*saxum*) and an Indian double-headed string-tensioned conical drum (*dhol*).[8] Who, I wondered, gets to name the crater that DART hopes to leave behind? Stickle looked to the public relations team standing off to the side, and they shrugged. Nobody had thought of that yet. "I imagine you have to see it to name it," she said—which meant it would probably come down to Patrick Michel, the commander of the postimpact Hera mission.

I had absolutely no idea how Cristina Thomas, the woman wielding the world's telescopes, managed to speak coherently just a few hours before impact. How many observatories—including the Hubble Space Telescope, the Lucy spacecraft, and the James Webb Space Telescope way above our heads—did she have under her reign at this point? "I've honestly stopped counting," she said. Some reporters intimated that their editors would need a more precise figure. "I don't know anymore, because there's a lot of them. It's incredibly exciting to have lost count," she said. But if you had to put a number on it? Probably more than thirty-six observatories, with one on almost every continent—all of which would zero in on Dimorphos just prior to impact, and keep watching it until six months post-collision, as the injured asteroid and its cloud of wreckage evolved.

It wouldn't take too long to get an idea of how much the impact altered Dimorphos's voyage around Didymos—they would know in a few weeks, maybe even a few days, she said. The light reflecting off Dimorphos would let them track how long it took for it to slip behind then emerge from Didymos during its orbit, something that radar observations would eventually shore up with greater accuracy.

After the press briefings, everyone shuffled into the main atrium, wondered where the coffee was—not present, it grimly turned out—and waited. Our soundtrack was an incongruous blend of the voices emanating from the NASA TV show on the screens above us and the staccato clacking of a hundred laptop keyboards. The countdown kept ticking down: three hours to go.

An announcement: the impact should happen at 7:14 p.m. local time. It was now 6 p.m. We all sat up in our chairs, eyes on the screens. Savitsky, sitting next to me, looked around at the other journalists, then leaned in to quietly offer an idea: we should try to find out where the DART team members are. They had clearly been moved out the way of the press, partly so they could focus on what they are doing, and partly so they could enjoy the event as it unfolded. But they probably wouldn't mind two science journalists becoming flies on the wall, right?

It was 6:06 p.m. I caught a few people pointing at the screens in my peripheral vision. DRACO was sending back images of Didymos, just a few pixels wide, a fuzzy ghost drifting through space—and it was expected that, at any moment, just to its right, a single pixel, distinct from that specter, would swim into view: Dimorphos. At this stage, it was still invisible, but anticipating its emergence played tricks on our minds. Marina Koren, of the *Atlantic*, said she kept mistaking specks of dust on her laptop screen for the elusive asteroid—a perfect description of DART's tiny target, now just 14,000 miles away.

Separately, Savitsky and I messaged a few DART team members to inquire as to their whereabouts. Two respondents said that they were across the campus, about a twenty-minute walk away, and that it would be fun to see us if we managed to sneak in.

We thought about it for a second. If we made it to the mission control room, or at least somewhere close with all the DART team present, we would both capture the raw emotion of the moment, not the postimpact curated presser that everyone else would go to. But if we got lost, or the media relations team caught us somewhere we technically were not permitted to be, we might get sent back to the main atrium and miss the main event en route, which would be a disaster.

As we weighed our options, the clock ticked over to 6:14 p.m., and Adams was seen on the TV screens running a systems check in the control room. People at their stations nodded or audibly confirmed that everything was looking marvelous. "We're ready to burn," said one of them, awesomely. "Autonomy is nominal," announced someone else. The spacecraft's connection to Earth's Deep Space Network got the thumbs up: we're speaking to DART. Adams chuckled. "It looks great," she said, staring down a newly resolved, unsuspecting pixel. "We are starting to see Dimorphos for the first time. It looks great!"

We had to decide. So, we did—we booked it across campus. I put an earphone in, listening to the NASA TV livestream for any updates.

The Applied Physics Laboratory has a weird layout, with similar-looking buildings spread out just far enough that it is difficult to tell one from the other, scattered across a campus that intersects a busy road. We picked a direction that felt right and started walking.* Eventually, we began to hear a ruckus to our left, across the road and on a part of campus shielded from view. It sounded like a street party. In my ear, I heard Adams conducting another systems check in the mission control room. DRACO was solid, SMART Nav was solid, autonomy was looking good, so were communications. I checked the time: 6:46 p.m., close to the point of no return.

A bunch of security guards stood at the entrance to whatever it was

* As we passed one building, a limousine parked itself outside the entrance, before several besuited people with aviators stepped outside the car and opened the door. In our haste, we almost tripped over the VIP that left the vehicle: Bill Nye the Science Guy, on his way to give a private talk with some of the DART team, by the look of things. There was no mistaking that bow tie.

that was transpiring around the corner. They caught our eye, and we nodded and strode confidently forward. We had made it through, but to what?

As we turned the corner, all was revealed: what looked like every single member of the DART team was there, plus their families and friends, and probably every single person who worked at the Applied Physics Laboratory. Large tents were set up across a vast area, filled with drinks, delicious-looking food, and—was that coffee? We each took a cold beer. We had earned it. Someone who I think was an astronaut was giving a talk to a packed crowd under one gazebo. TV screens streaming DRACO's imagery were everywhere. As we explored the campus, we kept an eye out for anyone we knew, but there were hundreds and hundreds of people swarming one way or the other. At one point, Spot—the autonomous dog-like robot built by Boston Dynamics—sauntered past us.

It was 6:49 p.m. Where was the mission control room? We didn't quite know where to go, so we aimed for a set of big doors that had a river of people streaming in and out. Down the hallway was a cavernous room, filled with about 200 scientists, engineers, and, crucially, massive screens on every wall. It was 6:52 p.m. By then, it was extremely unlikely we would find our way into the mission control room itself—and besides, we didn't want to be a distraction to Adams and her team. This auditorium looked perfect.

After politely squeezing to the front, we stood among speechless stargazers, their jaws dropping as Didymos began to fill up the screens, bit by bit. This asteroid—a world of silvery debris, completely motionless at some moments and breathtakingly kinetic at others—loomed ever larger—as did Dimorphos, which began to transmogrify from a small gray smudge to something more tangible. Adams and her team were holding on to their seats for dear life, or clasping their hands together, eyes unblinking. Right at that very moment, SMART Nav's objective was to spy Dimorphos materializing in the corner of

DRACO's eye, and switch from targeting Didymos to locking on to Dimorphos.

At 6:54 p.m., a voice over the livestream declared that we had precision lock. DART was now focusing exclusively on Dimorphos. The trench run had begun. The cathedral full of scientists became, for a minute, slightly less tense. People around me were excited by the sight of this brave new world. A spacecraft was about to fly extremely close to an asteroid at 14,000 miles per hour and head directly to another one, and we were watching it live as it happened. This was science fiction made real.

"It's the final countdown," someone behind me said, before letting a giggle escape into the room. Someone else added, "Will things ever be the same again?" And that was it: "The Final Countdown," the exceedingly apposite track by the band Europe, was bouncing around my skull as I watched a spacecraft try to achieve something preposterous. At 7:10 p.m., Didymos started to slip out of sight—and all eyes, on Earth and in space, were on Dimorphos, just 920 miles away.

The room hushed. Giddiness was overwhelmed by wonder. Dimorphos drifted into the center of the screen and rapidly inflated like an interplanetary balloon. Boulders, valleys, cliffs, and chasms emerged; some were bathed in starlight, while others hid in the darkest shadows.

It's 7:12 p.m. DART looked like it was piloting itself perfectly—but it was difficult to tell from where we were standing. It could still miss. What if, after all this time, it did miss? It was too late to change anything now. The hands were off the wheel, and no one in mission control could initiate any last-second maneuvers. Someone in the room shouted: "Two minutes!" and the tension skyrocketed, both in the packed room and in mission control, whose attendees could be seen on a pop-up screen in the corner of DRACO's view of deep space. Some people were as rigid as stone; others vacillated as if affected by an earthquake. Someone next to me suddenly crushed their not-quite-empty plastic cup of coffee, the lukewarm liquid spilling over their hand.

Dimorphos was now 460 miles away; its geologic subtleties and ornaments were starting to become visible. "There's no way that's a solid rock," someone declared. Dimorphos was a rubble pile, not much different from the awkward, weakly bound asteroids visited by Hayabusa2 and OSIRIS-REx. DART was not going to gently land on the surface of Dimorphos, but ram it at an incredible speed. If it did impact the asteroid, would it hit it too hard?

7:13 p.m. One minute left. "Holy shit. Shit shit shit," uttered someone to my left. The hum in the room rose to a rumble. About thirty seconds to go. It looked like it was going to happen—would it? If it had, the spacecraft was already dead, and we were watching its memories. But Dimorphos is so small, maybe DART just barely missed it? For everyone in that room, this was the longest thirty seconds of their lives. In mission control, Adams—front and center—began counting down from ten. Nine. Eight. Seven.

People began to shout and scream. Dimorphos rushed at DRACO until the white-gray realm was all anyone could see. I was in an ocean of noise. "Oh my god," someone said. Adams began to leap into the air—and the screen went red.

Loss of signal.

When I was nine years old, I saw something in the sky that's hard to adequately describe. I mostly remember what it felt like: I was happy, certainly, but I also knew that the world—the universe, really—wasn't what I had thought it to be. It was considerably more spellbinding. Reading about comets, and seeing photographs of them, was one thing—but seeing one up close, one so luminous and majestic, was transcendental.

It was the spring of 1997, and I was on the back of a canal boat on a vacation somewhere in England with my family. At that moment, not long after the twilight had dissolved into night, my mum was somewhere else on the boat, negotiating dinner options with my grandparents. My dad was standing next to me on the back of the boat, and we were both looking skyward, watching Comet Hale-Bopp and its ethereal, twin tails—one a coconut white, another ultramarine—light up the world.

Discovered just two years earlier by astronomers Alan Hale and Thomas Bopp, this comet turned out to be one of the brightest comets to grace the inner solar system in recorded history.[9] It was 120 million miles away from Earth, but it looked like an alabaster explosion. Its nucleus was thirty-seven miles wide, a snowball of erupting ices. I didn't really know what to say, and neither did my dad. What could you say when confronted with such staggering beauty?

"Wow," I said. "That's really cool." My dad let out a laugh, the same laugh that has punctuated so many important moments in my life.

"Yes, it is," he said, patting me on the back. "It really is."

The cosmos no longer felt static. Everything seemed like it was in motion, each object an instrument in a galactic orchestra, all combining to make something wonderful. But space itself never felt farther away. It was filled with behemoths, titanic creatures without thought, all making their way through the stars without any barriers. They may pass by our world, but they don't stop to have a conversation. All of them—planets, moons, asteroids, and comets—just keep going. I knew astronauts had set foot on the Moon long before I was born, but the universe was always going to be as it is now. It was a place that affects us, never the other way around. To change that would require nothing short of magic.

September 26, 2022, 7:14 p.m. There it was, the last shot of Dimorphos, taken by DRACO just a few miles above the surface, as if it were a bird hovering calmly above, looking for prey. The next screenshot was bright red—violence had happened. It actually bloody well happened!

The loss of signal announcement confirmed it: DART impacted Dimorphos, almost directly in the middle—a perfect bullseye.

Everyone was jumping up and down. I started cheering myself; Savitsky, grinning madly, emerged from the tumult. Fists were pumping into the air; there was cheering, crying, shouting coming from almost everyone, save the handful of viewers standing still, eyes wide, mouths agape, reeling in giddy disbelief. Those in the mission control room were wildly applauding, and Adams looked like she was going to melt from the ecstasy of it all.

"That was a hell of a thing," someone told me. "A hell of a thing!" Another scientist, almost unable to stand up, looked me dead in the eye and said, quite justifiably, "holy fucking shit," before wandering off, drunk with joy. "That was dead center, man, oh my god," said another, while their friend put their hands to their face, saying: "I have a career now!" For others, this would be the capstone of their careers. And for DART, it was over—its trail of debris, yet unseen, was now being scattered like glitter across the darkness, a self-made elegy for a triumphant moment in our species' history. The stars had surely been shifted, all through the sorcery of science. And I couldn't help thinking the same thing, over and over: everybody wins—just this once, everybody wins.

XI

Dust to Dust

After DART's rousing success, nobody really knew what to do with themselves—so everyone oozed outside into the humid night. Celebratory drinks began to emerge from seemingly nowhere. I came across Kelly Fast of the Planetary Defense Coordination Office, who was pointing excitedly at something on her phone.

She showed me an emailed GIF that had come through just a few moments ago. It showed a blob moving right to left, before suddenly getting brighter and ejecting an enormous arc of luminous matter into space. It was Dimorphos at the moment of impact. That meant two things: one, Cristina Thomas's master plan of global observations was already working; and two, DART really rang that asteroid's bell. It looked like a lot of debris—maybe a bit too much.

The observations used in the GIF were made by Amanda Sickafoose at the Planetary Science Institute, who was watching Dimorphos with a telescope in Sutherland, South Africa.[1] Sickafoose was one of the first astronomers in the world to witness DART's spectacular death, so I had to call her as soon as possible. "I'm not an asteroid person," she told me, explaining that she mostly observes things that live in the outer solar system. But she's good friends with Andy Rivkin, and he helped convince her to join the mission because her position in South Africa meant that she could get first dibs on the impact imagery. Just before the fireworks, she and her colleagues sat with two laptops open. "One had images of the spacecraft coming in," she said, and the other showed real-time images from the telescope trained on Dimorphos. "And I was so pessimistic." She thought Dimorphos would brighten just a little. Her colleague, Nicolas Erasmus, was more optimistic. "No no no, it's not going to be like that," she told him.

Sickafoose was thrilled to be proven wrong. The moment the DRACO feed went red, they turned to watch the telescope's perspective. "It just got brighter and brighter and brighter. And then that bit splits off. It was absolutely extraordinary," she recalled, laughing. "In the first few minutes, we were all watching it in awe, and amazement, and wonder. But immediately it was like, my gosh, how big is that scale?" Another telescope on-site—part of ATLAS (Asteroid Terrestrial-impact Last Alert System)—captured a similar series of images. The amount of debris the impact appeared to have created was almost excessive.[2] It remained to be seen what exactly this meant for the deflection campaign—but at the time, everyone was simply relieved that DART had hit its target.

I couldn't find Andy Rivkin among the thrilled and exhausted planetary defense revelers at first, so I sent him a message. "The experience has been like nothing else I've ever had," he responded. "Thrilling on an emotional and intellectual level. Wonderful to share with the team. And yes, now we get to work."

There was one last press conference to go before everything was shut down, so Savitsky and I slowly made our way over, still buzzing from the entire thing—and trying to, but quickly giving up on, inventing a cover story for anyone curious as to why we had vanished. Just before we departed for the other side of campus, I snuck off for the briefest of bathroom breaks. As I was leaving, three scientists, arms around each other, walked past me in the corridor, tipsily singing the lyrics to the chorus of Aerosmith's *Armageddon* classic, "I Don't Want to Miss a Thing." It's not full-blown karaoke, I thought. But it's good enough.

Eventually, all the journalists, camera crews, and photographers were gathered in an auditorium, awaiting the day's final words from a select few—including Adams, the chieftain of DART's human copilots, who was as mentally elated as she was physically deflated. The impact was clearly successful, but when will we know if the deflection worked and the orbit of Dimorphos had been changed? A few days, perhaps a week or two, she said. The LICIACube images were being downloaded at that very moment, she added; hopefully it managed to capture the impact on camera. A follow up: should all Earthlings rest a little easier tonight? "As far as we can tell, our first planetary defense test was a success," she confirmed. The packed auditorium full of reporters, scientists, and press officers erupted into cheers and applause. "Yeah. I think Earthlings should sleep better," Adams said. "I will."

October 14, 2022. I was sitting at Flagstaff's Lowell Observatory, surrounded by coppery leaves. Teddy Kareta—DART team member and my tour guide to Meteor Crater—had his laptop open, and he was showing me what I consider to be one of the most inspirational images ever taken of anything in space: the SOAR Telescope portrait of the impact's aftermath.[3]

Back on September 28, Kareta had teamed up with fellow astronomer Matthew Knight at the US Naval Academy; they remotely hijacked the Southern Astrophysical Research (SOAR) Telescope in Chile, and used it to soak up the light bouncing around the debris left behind by DART's demise. What they captured surprised and delighted everyone. Around a core of the whitest light is a diffuse rainbow of reds, greens, and blues; emerging from that core are multiple linear tendrils of evanescent, illuminated matter. The longest, the streak flaring out toward the top-right corner of the image, was at least 6,200 miles long.

It was debris bleeding away from Dimorphos, an asteroid enveloped in blindingly bright reflected starlight. DART impacted it so powerfully that it transformed from an asteroid into a pseudo-comet with an extremely long tail, one that emerged instantly after the spacecraft's angry rendezvous.

Like everyone else in the community, Kareta was clearly blown away by the SOAR image. They had, after all, documented humanity accidently making a rocky comet. And this image quickly found its way around the world, onto the printed and digital pages of dozens of magazines and newspapers both tiny and mighty. "Millions of people have seen the image we made. Or rather, we took the data for it and. . . ." he paused. Was the amount of exposure this single image received daunting? "Yes!" he exclaimed.

Kareta laughed. "I'm really happy people saw it. This is going to be one of their few points of access into the mission. They'll have heard that DART existed, they will have heard that DART hit the thing, they'll see the picture, and probably they'll have heard the news—that DART really fucking worked," he said, almost slamming his fist down onto the table. "I tried to provide that quote to the *New York Times,* but they didn't want to print it."

And then there was LICIACube, coming in clutch. Nobody quite knew how effective of a videographer this toaster-size satellite would be, but it was a relatively inexpensive bit of equipment that could hopefully capture the image of a lifetime, so it was always worth a try. The robotic paparazzo zipped past the impacted asteroid at 22,000 miles per hour, just three minutes after DART crashed into it. Although some manual piloting commands were sent to it initially, the ground team had to rely on the LICIACube's autonomous tracking software to get the job done. Did it work?

"About one hour after we had the communication window with the antennas on the Earth, we were waiting for the first data to come," Simone Pirrotta, the CubeSat's project manager at the Italian Space

Agency, told me. "When the first images appeared on screen, it was super crazy. It was our moment to celebrate, and drink."

The images will likely be as iconic to the mission, and the history of planetary defense, as the SOAR shot. Around a bright nucleus was a spider of light, a fuzzy set of arachnoidian limbs reaching around Dimorphos, as if it were being ensnared by an eldritch monster older than the stars themselves. It is an astounding, evocative image—and one that briefly caused concern. Upon first seeing it, scientists wondered if DART had hit Dimorphos a little too zealously.

One such fretter was Patrick Michel, the head of the Hera mission from the European Space Agency. He had also been on campus during the impact. For him, the DART mission's climactic moment "was a mixture of many feelings," he told me. Joy, certainly, after watching the result of decades of effort—but there was also frustration. "In principle, ESA should have been there at the same time," he said, lamenting the delay that meant their mission would arrive a few years post-impact. "Otherwise, it was absolutely crazy." There was also anxiety, he explained. Dimorphos was a rubble pile, which meant it was weakly bound together. If you hit it hard enough, you wouldn't deflect it—you would disrupt it, fragmenting it into many pieces, some of which may end up on potentially dangerous orbits around the inner solar system. Even if the freshly made shards turned out to be harmless, it would have left Hera without an asteroid to visit, robbing it of purpose.

According to the simulations of the impact run beforehand, though, disruption was not a plausible outcome. "We didn't find any cases where we fully destroyed Dimorphos," Michel recounted. "But. Hm. When we first saw the images by LICIACube, some of us really thought that we destroyed it. Including me."

The whole point of the DART mission was to prove that deflecting an asteroid with a spacecraft was possible. Annihilating it would be an all-time whoopsie. "When we started getting the first immediate results back, many of us were a little bit concerned that we might have disrupted the asteroid," impact simulation queen Angela Stickle told me, over another call.

Upon seeing those initial LICIACube images, which came in just a few hours after DART's destruction, "we were not very calm," Michel said. Mercifully, succor arrived one day later: radar managed to ping off Dimorphos, proving it was still there and mostly intact. That must have been an interesting twenty-four hours, I said. "Mm." Michel looked unamused.

What mattered most, though, was that Dimorphos didn't just inertly absorb the strike. It was deflected. Its original orbit around Didymos took 11 hours 55 minutes. The DART team hoped to reduce that by just over a minute. But early observations revealed that they had almost hilariously overshot that low bar. Dimorphos had been knocked so much closer to Didymos by the impact that its orbit was now 11 hours 23 minutes—a reduction of 32 minutes.[4] This was no glancing blow; DART's sucker punch landed right in the center of the asteroid—a perfectly aimed, emphatic impact. The mission team's campaign wasn't just a success. With flying colors, they showed that Earth could be defended by a pocket-size spacecraft.

After managing to peel myself away from that remarkable SOAR image of the comet-like Dimorphos, I headed down from the observatory, walked across Flagstaff, and soon found myself inside Cristina Thomas's office at Northern Arizona University. Books surrounded me, as did posters for various interplanetary robotic adventures of yore. The sun was beaming, as was the coordinator for the world's Dimorphos-tracking observatories. "This was the first space mission I was seriously involved in," she said. Could've fooled me—she pulled off her part without a single major operational hitch. "This entire thing actually worked." Every observatory pointed itself in the right direction at the right time, and for many of them, the weather was beautifully clear.

I had last seen Thomas—very briefly—back in the sea of people outside on the Applied Physics Laboratory's campus, standing next to Kelly Fast in front of one of the TV screens. How was her experience? "There's screaming everywhere. There are fireworks. It was phenomenally exciting," she said. Perhaps an hour or so afterward, images from South Africa popped up on her phone, showing an extremely bright flash of light emerging from the asteroid. "Whoa," she thought. "What's happening?"

It turned out that the initial ejection of matter recorded by those telescopes wasn't rocky debris. It was more likely to have been xenon and hydrazine fuel—the stuff DART used to navigate through space—turning into vapor upon impact, violently expanding, then rapidly cooling and condensing into a visible, arc-like cloud. But that illusive outburst previewed the spacecraft's genuinely destructive performance. Follow-up observations, like the SOAR and LICIACube shots, confirmed that Dimorphos had indeed been strenuously walloped.

Inundated by hundreds of images taken by her legion of telescopes both on and off the world, all of which showcased this incredible victory over orbital physics, Thomas was thrilled but also in a state of denial. That the mission happened at all was inherently shocking. "[NASA] said: 'We're going to make this, this is actually going to happen,'" she told me. "I sort of expected them to pull the rug out from under us." It would be a long time, another generation perhaps, before another mission would have the scientific, political, social, and literal impact that DART had. "It harkens back to these '90s sci-fi ideas that people had, you know?"

So I dared to ask: What's next? NEO Surveyor, please and thank you, Thomas replied. "We're way behind on this congressional mandate. Embarrassingly behind." When you tell people we've found less than half of the city killer NEOs, there's this "record scratch moment." Finding the remainder with both space-based and ground-based observatories, she said, "has to be the next big thing." The hope is that America doesn't get distracted by other space shenanigans.

DART was a steal at $314 million. Thanks to the unnecessary hoops it was forced to jump through, NEO Surveyor has become inordinately expensive. But finding almost every single potentially lethal, city killing or country crushing asteroid near Earth for just $1.2 billion remains the bargain of the century. The entire planetary defense community agrees that NEO Surveyor needs to happen. But there is no guarantee that it will.

In November 2022, NASA announced that one of its two upcoming missions to Venus, the planet-mapping VERITAS spacecraft, would be put on indefinite hold.[5] This mission, one of the two winners of its most recent Discovery-class mission competition, supposedly had guaranteed funding. But NASA said that management and budgetary issues at the Jet Propulsion Laboratory had caused so much chaos that it was negatively impacting the VERITAS mission—something many in the space community deemed to be an unacceptable explanation.[6] Who's to say that the same won't happen for future planetary defense endeavors?

And then there is Artemis, America's effort to send astronauts back to the Moon and establish a permanent presence there (before China). Throughout the final stages of DART's life, NASA meanwhile tried to inaugurate the Artemis missions with a first test flight—sending one of its gigantic Space Launch System rockets and an empty capsule designed to carry astronauts into space—on four separate occasions. The first two attempts were scrubbed due to technical concerns, while a planned third was nixed due to the presence of Hurricane Ian. A rocket ultimately launched from Florida's Kennedy Space Center, and executed a perfect test flight, in November 2022.[7] Using NASA's own mega-size rockets (those based on increasingly outdated tech), the Artemis program was already five years behind schedule by the time of DART's death—and embarrassingly over budget at $40 billion,[8] a figure that is set to more than double by 2025.[9]

These lunar missions will indeed bring welcome scientific advancements, and many in the West would favor American (and European) dominance on the Moon and its valuable resources—water-ice to make rocket fuel and to support human habitats up there, for example—over a Chinese-controlled lunar surface. But the Artemis program is, for the foreseeable future, a politically motivated mission that puts America first. Planetary defense puts everyone first, and for a fraction of the cost. And currently, planetary defense is one-for-one on its very first deep space trial run. "Our budget is like two months of a delay for you," said Thomas, referring to the Artemis program. Comparatively, planetary defense spending is like a rounding error on a spreadsheet.

The hope, though, is that the momentum instigated by DART continues and never abates. "It's a thing now," she said. "We can bring people in."

The public can also move on from the *Armageddon* version of planetary defense, and truly begin to acknowledge that this is a legitimate career that people can pursue for the benefit of us all. "The NASA press folks almost seemed surprised at how jazzed people were about all this, about how many requests they got. They seemed incredibly overwhelmed responding to folks, to do interviews and stuff, because there were so many of them," said Thomas. I never understood the surprise: DART was a test of a technique that could save millions, perhaps billions, of lives one day. It's difficult to imagine a more aspirational, inspirational, exceedingly cool uncrewed space mission than that.

That doesn't mean DART and asteroid deflection translated perfectly to a general audience. Some people think, "We have hit something, and we don't know what we did. It could be on a crash course with Earth," said Thomas. Even now, after NASA has said almost a thousand times that Dimorphos posed no threat to Earth, that erroneous notion keeps coming up. "I've seen it more times than I feel good about. But maybe I'm spending too much time on Reddit," she said. "There were people who thought we were sending a bomb there." Some of her own family members asked her, "Why would you blow up the asteroid?" Seemed a bit excessive, they thought.

As I congratulated Thomas on pulling off her impact observation master plan, she dropped an unexpected bombshell. "We get to late August . . . and my grandmother died," she said. "And so . . . I got this phone call, it's like: you need to come right now. I drove out there." There was no way she could focus on DART, so she hoped that all the work she had already put into organizing everything would be enough. "I was able to go out there before she fell asleep." Most of her colleagues were unaware that she had just suffered a heart-splintering loss—and that she had helped organize the funeral—because everything she was involved with on the DART mission was working perfectly. "It was absolutely fine, which was kind of shocking to me, because I was very all over the place," she said.

Did Thomas's grandmother know what she was working on at the time? "For the most part," she said. She watched her granddaughter's appearance on local TV during a segment that aired when DART launched, "which she was totally jazzed about." She also stayed up to watch every single edited version of these interview segments, even if the changes each time were minimal. The small smile Thomas had worn for the last minute or so then faded away. "I was kinda looking forward to telling her how it went," she said.

Thomas wasn't the only mission team member in mourning. Not long before DART's impact day, I sent Jessica Sunshine—the asteroids, comets, and impacts expert on the team who also played a key role on the Deep Impact mission—an email, a journalistic request for an interview. I didn't get a reply until October 18 for a nightmarish reason: her mother had passed away just a day after DART hit Dimorphos. "Not surprisingly, I'm a bit overwhelmed dealing with both; the largest range of emotions imaginable," she wrote back. "But if you're not in a huge hurry, I'd love to talk."

I insisted that we need not conduct that interview, but I'm glad we ended up connecting. Like Thomas, Sunshine waxed lyrical about the DART mission despite her recent, terrible loss. "For drama, I don't think you can exceed an impact. I'm lucky enough to have two in my life," she said, referring to her time on the Deep Impact mission. To her mind, nothing beats space exploration. "It's like a drug. When a world goes from a dot... when it goes from an astronomical target to a geologic world, to me, is the most thrilling thing you could ever do."

The innate coolness of DART will take some beating—as will its inherent weirdness. "This is a strange mission in that... it was essentially a camera," she said. And despite that simplicity, "we didn't know how or if it was gonna work." That they were allowed to try at all was what mattered. "It was something that needed to be done, and it needed to be proven."

Like many on the DART team, Sunshine was now cheering on NEO Surveyor from the stands. "We can bring the risk down so far by doing the two things we've always talked about: mitigation and detection. You really can't say that about anything else, natural hazards-wise. If we could end the threat of nuclear weapons, we'd do it," she said. As socially conscious scientists, defending the planet from asteroids is a major responsibility. To her, there was no deliberating the matter. They had no choice. "This is something we have to do." And by "we," she meant everybody: "This is not a US problem. And it's not just the US that can contribute." DART was a huge international effort, as Thomas's observations program showed, and as Hera will demonstrate in the coming years. "That's just tremendous."

Sunshine's aspiration—that all spacefaring nations should be impelled to act—felt ethically bulletproof. DART's explosive triumph was a proclamation: with a little more effort, we can permanently solve a global problem in a relatively straightforward fashion. The familial losses experienced by Thomas and Sunshine on either side of the spacecraft's grand finale emphasized the novelty of this endeavor. Disasters—from powerful earthquakes to pandemics—can be mitigated to varying extents, but the threats themselves, and some of their lethality, cannot be eliminated. We can make homes earthquake resistant. We can raise barricades against tsunamis. We can get vaccinated. But on an individual level, there is often nothing we can do to guarantee that death won't pay us, or those around us, an untimely visit.

But planetary defense—that's different. A sizable asteroid will hit the world, somewhere, someday, if we decide to do nothing about it. But DART demonstrated that we exist on a timeline in which we are choosing to prevent this from happening. Scientists and engineers are scheming to save the lives of not just those they know, but those they will never know—simply because they can.

No matter where the planetary defense effort goes next, Sunshine emphasized that everyone already owes a lot to Lindley Johnson. "He's really smart. He knows what he's doing. He's quietly stubborn. He just doesn't go away," she said. "He doesn't piss people off."

Did the maestro himself have any notion of where we might go next? I phoned Johnson to ask. "We can't do anything about 'em until we find 'em," said Johnson—another vocal supporter of NEO Surveyor. "I thought we could have done that ten years ago."

The Planetary Defense Coordination Office's Kelly Fast, also on the call, agreed. It was crystal clear that they could have realized NEO Surveyor sooner if the resources had been efficiently afforded to them. "I hope we don't have that regret where we discover something one day that's a problem, and we go: oh my goodness, we would have known about this earlier," she said.

Johnson expressed his support for testing out another planetary defense technique—a gravity tractor, perhaps, or ion beams. "DART certainly confirmed that the technique of a kinetic impactor does sort of what we expected it to do," he said. But each asteroid is going to be different in some way, and a kinetic impactor won't work on all asteroids, especially the larger ones. "It is also a little bit of a blunt technique," Johnson noted.

But what about the funding for the PDCO? Getting $150 to $190 million per year is still chump change compared to NASA's total budget, let alone America's sum annual expenditure. Johnson, though, did not suggest they needed a vast injection of funding. His office will continually work out what financing is required to shield the world from killer asteroids, and he hopes that the higher-ups at NASA, and those in Congress, will always grant those requests.

But the future of the PDCO is not set in stone—and who knows what decisions future administrations will make? Either way, it seems likely that an increasing number of countries will start throwing their hats into the planetary defense arena. Details were sparse, but China had recently announced that it, too, was going to deflect an asteroid with a spacecraft in the coming years,[10] hoping to show that America isn't the only nation capable of defending Earth from extraterrestrial threats.

Johnson told me that he was thinking a little more about his retirement these days—fair enough, I thought; DART would make an excellent career capstone. But he stressed that everyone must remember that DART is the beginning of a safer world, not the last word in planetary defense. "It ain't over yet," he said. "There's still work to be done. Things to hand off to the next generation. We've moved things along, and we've got it at a point now where it will continue. It won't die, just because I retire."

December 15, 2022. I was in Chicago at the annual meeting of the American Geophysical Union, the world's most populous gathering of Earth, space, and planetary scientists. Some of the DART team was there—including Andy Cheng, Andy Rivkin, Megan Bruck Syal, Angela Stickle, and Cristina Thomas—sharing the latest results from their postimpact analysis.

That comet-like tail was still there, several months after the impact. Rivkin pointed to an image of it on a large screen. "It's stretches off the edge of the frame. It's tens of thousands of kilometers long," he said. "We're thinking maybe 30,000 kilometers, maybe longer." (That's nearly 19,000 miles.) So how much debris was thrown into space by the impact? "We think it's about a million kilograms of material, at the least," he said—about 1,100 imperial tons, or if you're the *Jerusalem Post*'s legendary Aaron Reich, almost ten blue whales' worth of matter.

Andy Cheng then took to the stage to discuss how they had been measuring the deflection campaign's success.[11] Shrinking the orbit of Dimorphos around Didymos from 11 hours 55 minutes to 11 hours 23 minutes had already proven that asteroid deflection worked. But how much, if at all, did DART punch above its weight?

When DART hit Dimorphos, it transferred its momentum to the asteroid. Technically, momentum is the mass of an object multiplied by its speed, but you could also describe it as the oomph something has when it hits something else. If you throw a brick at a window very slowly, it may not have enough momentum to smash the glass. Throw that same brick faster, and you have given it more momentum—enough, perhaps, to shatter the pane.

Scientists wanted to know what happened with DART's momentum. It was obviously transferred to the asteroid, but it may not have been a one-to-one transferal. Sometimes, the physical nature of an object means that if you hit it with something, it flies backward as if given extra momentum. This may have happened with Dimorphos: so much material flew off it during the impact that it may have acted like a rocket booster, giving the asteroid more thrust than the impact alone would have imparted.

According to Cheng, this is exactly what happened. Beta, a scientific metric that measures the momentum enhancement of an impact event, came out at 3.6. That means Dimorphos was knocked back as if it were hit by 3.6 DART spacecraft all at once, thanks to all that thrust-making debris that flew off it. This result means that, for city killer asteroids that are often rubble piles, a DART-like deflection mission will likely push that asteroid away more effectively than anyone expected. "Suppose you had a situation where an asteroid is coming for Earth," Cheng explained. With beta value of 1.0, you could get it to barely miss the Earth if you hit it eighteen years prior to impact. "If the momentum enhancement factor is 3.6, you could use that same kinetic impactor and the same target, and you could do your interception five years out, instead of eighteen years out. You'd need less warning time." And if the asteroid was heftier, you would be able to deflect it with a smaller spacecraft than originally thought prior to the DART mission.

In other words, deflecting an asteroid with a spacecraft, although not easy, was an even better defensive technique than its proponents had previously considered. "If you're trying to save the Earth, that makes a big difference," said Cheng.

At the end of the press conference, all the DART team members who were present gathered for a big group photo at the front. One journalist shouted out: How do you feel? Cheng, not missing a beat, shouted back: "We feel great!," to a round of applause.

Punching an asteroid is not what you would call an exact science. Everyone suspected that Dimorphos's orbit around Didymos would shift—but precisely how Dimorphos would respond to this act of vandalism was always uncertain. How would a swarm of rocks flying in close formation react to a head-on collision with a speeding robot? As it turns out, one could answer this query with a single word: flamboyantly.

By the time 2023 dawned, Cristina Thomas's observation campaign had shown beyond all doubt that the deflection campaign had been a success. But it had also revealed that Dimorphos was still reeling from the impact. For a few weeks, a second debris tail could be seen streaming away from the asteroid; it seemed likely that DART's impact had liberated a sizable shard from it, one that eventually returned home with a bang, creating a temporary river of ruin.[12] Astronomers outside the DART team also got involved; they pointed the Hubble Space Telescope at Dimorphos, and found that it was surrounded by a swarm of house-size boulders—wreckage that had been gradually peeling off the bruised asteroid during the winter months.[13] None of these observations were especially strange—more interesting than surprising. But a shift in the amount of light the asteroid was now reflecting suggested something genuinely bizarre had also happened: the entire asteroid had changed shape.

Sure, Dimorphos was always going to transform a little if a spacecraft was going to hit it so hard that some of its material would be blasted away into the void. But the ejection of matter couldn't fully explain what they were seeing. Pre-impact, Dimorphos was an oblate spheroid—a largely spherical object flattened a bit at the poles, like Earth. But now, it appeared to be an elongated ellipsoid—an oval. Everyone had correctly guessed that Dimorphos was a rubble pile, but that it was quite this malleable was a marvel to behold. When DART crashed into it, the asteroid didn't respond like a brick wall or the side of a cliff; instead, it was as if NASA had fired a cannonball into a free-floating ball of mercury. The asteroid's surface may have splashed upward and outward. Some of those lithic droplets escaped into the beyond, but much of this fluid matter, trapped by the asteroid's gravity, curved around and collapsed back into itself. And when the waves eventually settled, Dimorphos was now an oval.

This only served to validate the experiences of DART's asteroid-prodding predecessors, like Hayabusa2 and OSIRIS-REx. By now, it was evident that these midsize space rocks were not remotely rigid or monolithic. They were not even mountains of loosely bound rocks, but something closer to amorphous seas of stone. Any future attempts to deflect these asteroids would have to take their potentially shapeshifting ability into account.

According to Angela Stickle, we should also expect DART's descendants to look and behave differently to the original. We last spoke several months after the theatrics of impact day, and she was still giddy from the emotional high it delivered. "I'm still trying to figure out exactly the right word to describe it," she said. "We all had goofy smiles. Our spacecraft worked, and it worked so beautifully. It hit where it was supposed to hit." The remarkable thing, she said, was that DART was basically a box with wings—not close to a perfect design for a kinetic impactor mission. "I wanted them to just put a big copper plate in the front, Deep Impact style, because it makes the cratering better." For a genuine asteroid deflection mission, you would also fill out the spacecraft so that there was no empty space, allowing it to couple its energy better to the surface of the target. "But one of the cool things that DART showed was that you don't have to do that," she told me. Even DART, a non-optimized deflection mission, worked wonders.

So, what's next? "Let's do DART 2. Let's go to another asteroid and smack it and see how much we learned." Sounds good to me, I said. And she would love to work on the sequel, too, if the original was anything to go by. "You think, I'm working on this thing to potentially save the planet," she said. "It's just cool."

Almost immediately after DART's death, Andy Rivkin waded into the vast sea of observations data—one whose depths he had barely explored by the time we spoke again later that winter. But much like Stickle, he also found himself frequently distracted by that dramatic day back in September. "It's been really something," he chortled. "It's just this amazing scientific thrill and this amazing emotional thrill."* Has he thought about the legacy that this mission is leaving behind?

---

* After the impact, Rivkin gave a series of press interviews, including one to the BBC. Just before he went live, he listened to the host giving the lead-in, introducing the next segment. "And I heard, 'And now, some good news from space!' And I thought that was awesome, that's wonderful. We're so glad. We're a feel-good story. That needn't have been the case."

Rivkin, someone with no semblance of a self-inflated ego, acknowledged that what the team accomplished was significant for everyone on the planet. But his response to this question was, as ever, thoroughly humble. One day, far into the future, history books will be written about what happened during the twenty-first century, he said: "And we're gonna be in the index."

And what of Elena Adams, the fixer of problems, the lead human copilot of the spacecraft? For her, the work ended the moment DART exploded into a million different pieces. How was she doing several months later? How did she feel about how it all went down? "I was really relieved, number one. There were many thoughts about what would happen if it didn't go as well." But "we hit within two meters of where we were aiming to hit. That... worked out pretty well," she said, with a smirk.* Now she was experiencing psychological whiplash. "The whole thing was fast. Fast to develop, fast to launch, fast to get there, and it was done."

How does one move on from something so spectacular? "One doesn't," she said, and laughed. A hint of sadness surfaced. "No, one doesn't, really. I've found it extremely hard to move on from that." None of her team really knew what to do. Some had moved on to other projects, but they had told her they were finding it hard to concentrate and engage with their new work. "I've had people in my office daily saying pretty much how hard it is to move on to other things. I'm definitely feeling that, too." Adams said that she was not currently working on a flight mission, so she felt especially rough. What's the greatest copilot in the solar system to do without a spacecraft to drive?

She was eagerly anticipating the upcoming Dragonfly mission, an Applied Physics Laboratory–led endeavor set to explore Titan, a muggy

* Their feat even made a segment on *The Late Show with Stephen Colbert*, during which the eponymous host respectfully referred to the team as "happy nerds." Adams thought about getting that citation printed onto some T-shirts.

and mysterious Saturnian moon. But the launch won't happen until 2028 at the earliest, and the spacecraft—a nuclear-powered quadcopter,[14] certainly a very cool design—won't reach Titan until the mid-2030s. "It'll be super exciting. But it doesn't really have that *moment*," she said, visibly frustrated. "There's no . . . big thing. It's gonna take nine years to get there, and then spend a long time exploring [Titan]. It's a science mission, there will be ups and downs, and it's extremely hard to do. But there's no . . ." she said, trailing off.

There is no substitute for practicing saving the world, I suppose—just as there is no replacing such a close-knit team, featuring some of the very best spacecraft makers and engineers on the planet. "These people have sacrificed a lot of themselves into doing this, especially during the pandemic. I just want to make sure they're not forgotten." I couldn't resist—somebody should probably write a book about all this, I suggested. "Well, get to it!" she said.

DART Vader, held by Elena Adams in DART's mission control room. *(Robin George Andrews)*

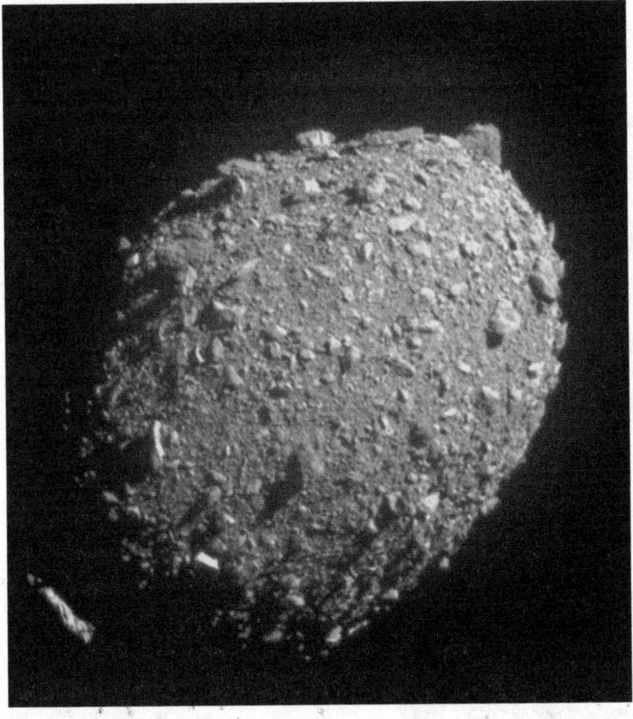

DART meets Dimorphos. *(NASA / Johns Hopkins APL)*

Elena Adams and her colleagues celebrate DART's demise. *(NASA / Johns Hopkins APL / Ed Whitman)*

Yours truly, celebrating with hundreds of scientists and engineers as DART hits Dimorphos. *(Zack Savitsky)*

This image from the Italian Space Agency's LICIACube shows the plumes of ejecta streaming from the Dimorphos asteroid after DART made impact with it. Each rectangle represents a different level of contrast in order to better see fine structure in the plumes. *(ASI / NASA / APL)*

Dimorphos's postimpact comet-like tail(s) captured by the SOAR Telescope. *(CTIO / NOIRLab / SOAR / NSF / AURA / T. Kareta [Lowell Observatory], M. Knight [US Naval Academy])*

The Hubble Space Telescope captures not just Dimorphos's comet-like debris tail, but multiple tiny specks all around it—boulders being thrown into space several months after DART's dramatic death. *(NASA/ESA)*

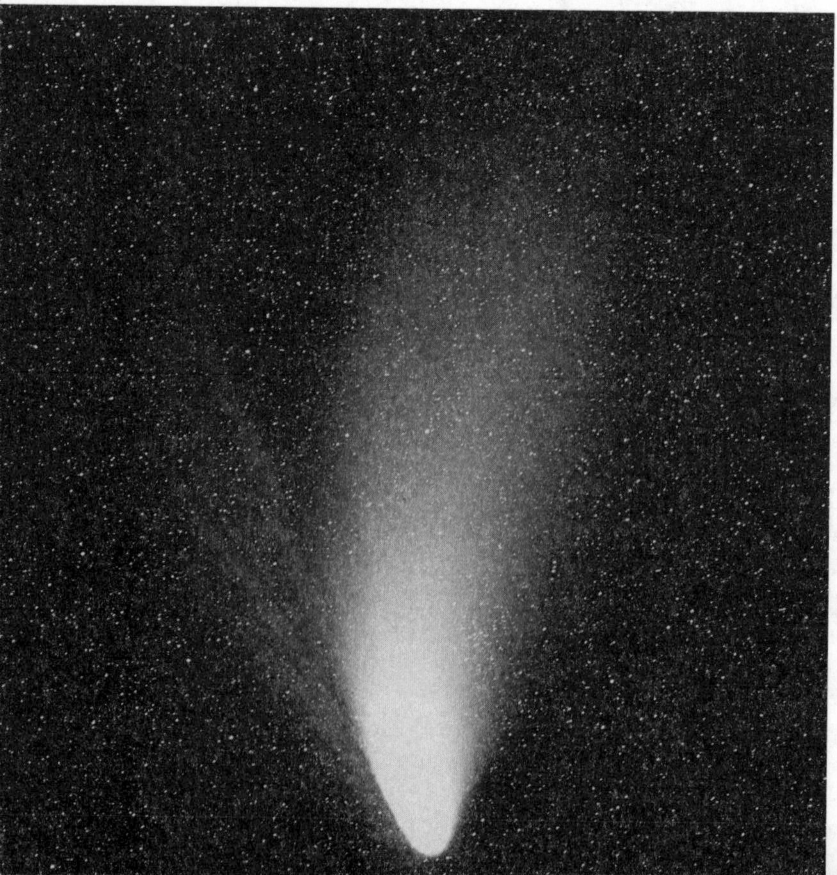

Comet Hale-Bopp. *(E. Kolmhofer, H. Raab; Johannes Kepler Observatory, Linz, Austria)*

EPILOGUE

What Do We Say to the God of Death?

Once upon a time, in the land of Termina, a lost soul came across a mask imbued with powerful magic—and, upon wearing it, made a terrible wish. That morning, the moon began to fall from the sky. It loomed ever larger above the world over the course of the next three days until it finally crashed into the ground. The land was scorched, the oceans boiled, and every living thing met with a terrible fate. All hope was extinguished—until a child from another land, a boy named Link with the ability to travel back through time, appeared just before dawn on the first day. Journeying across Termina, racing against the moon's descent, he found four giants, each imprisoned by a curse. After liberating these titans, the child climbed atop a clock tower, the site of the future ground zero. Using his ocarina to play a rousing tune, Link summoned the giants on the eve of the apocalypse. Standing around the clock tower, the four of them reached into the sky and held up the moon, stopping the world's end.

Many of you may not have heard of this myth until now—but if you are of a similar age to me, you may have immediately recognized it.

For millions around the world, this story was a childhood fairy tale, one that we did not passively absorb but actively played a role in. It is the narrative of *The Legend of Zelda: Majora's Mask*, the critically acclaimed, much-loved follow-up to *Ocarina of Time*, the legendary video game series' most revolutionary entry.

Majora's Mask is a fable steeped in sadness. The creative, exciting gameplay may have been primarily why so many of us enjoyed it way back in the year 2000. But its melancholic story is why it lingered in our minds as we grew into adulthood. Death is a ubiquitous presence throughout Link's odyssey, as are loneliness, helplessness, and grief—that is, until a recurring and seemingly unstoppable peril is irreversibly vanquished.

Ocarina of Time, a temporally chaotic adventure that features a volcano called Death Mountain, inspired me to become a volcanologist, then a science journalist—one who went on to write a book about how volcanoes allow us to explore worlds and times beyond our wildest imaginations. It is a coincidence that my second book is about a global effort to prevent asteroids impacting Earth. But it is no surprise that, as I wrote it, I increasingly dwelled on *Majora's Mask*. Perhaps you could think of NASA, ESA, and JAXA, and the countless observatories and universities across the planet, as metaphorical giants working to stall the moon's fall. What else is science if not a way to realize mythic feats? What is engineering if not crystallized magic?

Most of us do not work in planetary defense. But we have something in common with those researchers who do: curiosity, framed by compassion. Time is an unyielding river, one that pushes all of us downstream. No matter what we do, the river will win in the end. But we can do certain things to root ourselves more strongly to the riverbed—and to make sure others can hold on a little tighter, too. We can perform grand acts of philanthropy. We can choose careers that aim to make the world a more egalitarian place. But we don't have to dedicate our years to lofty humanitarian causes. All any of us *need* to do is empathize, with those we know and those we don't. Our individual acts of everyday kindness may not save the world. But our actions may save someone's life—or simply turn someone's downbeat day into something a little sunnier.

There are plenty of ways, both optimistic and pessimistic, to define human nature. For my money, preventing others from experiencing pain, loss, and grief is the most quintessentially human thing, the persistent idiosyncrasy that makes me think that, eventually, everything will be okay.

It's April 4, 2053. It doesn't have to be this particular date. I don't know when something like this will happen. But it could very well be April 4, 2053. So let's just say it is. And let's also say that somewhere out there, tens of millions of miles from home, a city killing asteroid is heading toward us.

In an alternate timeline, nobody would know of its existence, nor its voyage—but not this timeline. This 500-foot asteroid was identified years ago by an infrared observatory hanging silently among the stars. It was given a name: Izanami. It was then studied by every telescope on solid ground that was able to spy on it. And not long after, we sent an assassin by the name of Asuka to hunt it down.

On April 4, 2053, 9 billion humans are going about their daily business. Someone in Seattle is about to quit their awful job and leave their equally awful boss behind in the dust. A family in Edinburgh, walking into an animal shelter, is about to adopt that dog that they just couldn't stop thinking about since their last visit. A group of friends, celebrating a long-awaited reunion, are partying the night away in Seoul. And all around the world, a modest assortment of people are intermittently watching the NASA-JAXA livestream on their phones, or on TV; scientists are talking about Izanami, the asteroid that astronomers said stood a high chance of impacting Earth in 2098. Today is the day the uncrewed spacecraft Asuka springs its trap.

Nobody is especially concerned. The last time the space agencies tried this, using the kinetic impactor known as Rei, it went flawlessly. Asteroid deflection has become bizarrely routine. And this time, ESA's Heimdall reconnaissance spacecraft is also on the scene, following Izanami on its voyage, ready to document and examine Asuka's deflection attempt.

It's 10:33 a.m. eastern time. One minute remains. Izanami is barreling down its orbital racetrack, readying itself to make another close flyby of Venus. Seemingly nothing can stop its journey. But it's now 10:34 a.m., and the countdown ends; the Asuka impactor leaps from the darkness and plunges its copper nose cone into the asteroid's side. Heimdall sees a million shards of rock and metal cascading into the cosmos, creating a cloud of ruin so expansive it almost obscures the spacecraft's view of Izanami shifting off its preordained path. Back on Earth, thousands of scientists and engineers in several mission control rooms erupt into cheers. And 9 billion people continue to go about their ordinary, wonderful days.

I have no idea if, or when, something like this will come to pass. But I was thinking about this sort of scenario on January 24, 2023, while standing on a frozen lake close to the triple border of Finland, Sweden, and Norway. For my mum's seventieth birthday, I suggested that we should do something a little less orthodox than go out for a nice dinner to celebrate—and I was thrilled when my parents thought that adventuring above the Arctic Circle in the middle of winter was a most excellent idea.

It was certainly an efficient maker of memories. On that January day, we were zooming across an ice-dusted world on snowmobiles; my parents, pursuing me on their shared ride, were continuously taken aback by the astounding, Narnia-esque wonder of it all. At one point, the sky began to shimmer as the wintry Sun tumbled below the coniferous trees decorating the distant hilltops. Initially, reds and oranges emerged from the haze, like a regular sunset. But then the blue of the atmosphere gave way to a startling iridescence, as if a rainbow had melted and spilled across the sky.

We were witnessing something uncommon, a phenomenon created when high-altitude swarms of icy particles are bathed in sunlight from below the horizon, triggering an explosion of color.[1] These so-called nacreous clouds are infrequently seen above the Arctic Circle, and normally show up in small patches. We were fortunate that our sighting was especially vivid, something that proved so mind-blowing to our guide that she stopped seven times on our frigid voyage to photograph the sky, all while uttering various indecipherable expletives.

Earth is a great place to call home. Often, this truth is not self-evident; personal tragedies and global tribulations can easily convince us otherwise. But on that day, at that very moment, with my parents standing either side of me, I felt more sanguine than ever before. This world is beautiful. And it's filled with people who wish to protect it, and everyone on it—so that those who are yet to exist can one day look up at the sky and wonder how they, too, got so lucky.

ACKNOWLEDGMENTS

Writing this book has been the most fun I've ever had working on anything. If I went back in time and told my childhood self that, one day, I'd get to write a story about protecting the planet from a cosmic killer, I may have melted into an unrecoverable puddle of excitement and disbelief. My thanks go to anyone who has ever shown me a mote of warmth and compassion—but, naturally, a few of these altruistic souls deserve to be singled out.

To my inimitable editor, Helen, and everyone else at W. W. Norton & Company who thought that a book about saving the world for real was a good shout—you rock. To all the planetary defenders I conversed with, especially those on the DART team: you really did it, you utter legends. Thanks for the endless chats, for driving me around, showing me about, and trusting me with your tales. And to the media relations team and press officers who helped arrange so many of these vital encounters, particularly Ian O'Neill, Joshua Handal, and Justyna Surowiec: thank you for putting up with my thousands of emails. My gratitude must also go to Gene Roddenberry and George Lucas: without *Star Trek* and *Star Wars*, we may not have the nascent planetary defense system we have today.

To Zack Savitsky, whose journalistic determination literally led me to the emotional apotheosis at the book's climax: I owe you one, buddy.

To Victoria, one of the first editors to give yours truly a shot at doing proper science journalism, and who gave me a base in Washington, DC, along with her husband, Aalok: I tip my hat to you. To the science journalists Brian Resnick, Maya Wei-Haas, Michael Greshko, Nadia Drake, Issam Ahmed, Jonathan Amos, Annie Roth, Natalie Wolchover, and Marina Koren: your humor, kindness, empathy, and remarkable storytelling wizardry will always be a marvel to behold. Jennifer Leman, journalist extraordinaire: thank you for shaming me into flying out to Iceland to hunt down an erupting volcano with you; it was a blast. And, of course, an immense amount of gratitude to Ibrahim, for the wanderings, the world-sharing, and for making sure the birds don't go thirsty.

To my friends who graced my life before the authorial times: I'll never find the words to sufficiently describe your significance to me; instead, our story will continue to be written in laughter, tears, hugs, and adventures, for all the days to come, right until the ending of the world. To Lukas, Mercan, Johnny, Judith, Sebi, and Daniel; to Valentina and Andrew; to Alf, Chris, Jonny, Katy, Josh, Rosie, Tom, and Dani; to Kristy and Mariana; to Shobhit, Sonny, Geoff, Bryce, and Sandra; to Franzi and Kurtis; to George and the family; to Mitch and Simon; to Essie; to Sara and Kate; to Jason; and to Joe—thank you, all of you, for every single moment of every single day since we first met. Our shared memories are like postcards to the future, those that pop up in my mind to remind me just how lucky I am to have met you. I'm so proud of everything you have done, all you have accomplished, and everything you are.

Wondering where you are, Patrick? Don't worry—you get a special mention. It was a lifetime ago that we met at Imperial College during our first year of university, then almost immediately binge-watched *Neon Genesis Evangelion* and ran off to Paris to see Muse. By the time this book is out, we will have been on countless escapades all over the world, and you'll have been my best man at my wedding, flanked by fellow groomshumans Lukas, Sara, Joe, Geoff, and Valentina. I love you all, but Patrick: you're the original myth, the firing pistol on everything good that followed. Thank you, my dear friend. I hope my endless friendship is sufficient repayment.

Much love to the members of my far-too-populous extended family whose golden hearts are encased in platinum: Uncle Paul, Auntie Jackie and Uncle Yutaka, cousin Sean, cousin Neil, Lesley and the family, Franco, all the Szuliks—Mike, Caroline, Maisie, and Tommy—and John and Judith Nicholson. All my love, of course, to Nan Andrews, Nan Micallef, Nanu, and Granddad Charlie; I carry you with me, each and every day.

To those lost far too soon, taken away from us and the world so ruthlessly: Uncle Steve and Uncle Wayne—we keep you alive in our collective memory. To Rosa: the world was a little more magical and whimsical with you in it. To Mercan's family, lost to that awful earthquake and decades of corruption—Cahide and Rafi Sümbültepe (her aunt and uncle), Serhan, Başak, and Mahir Bozhüyük (her cousin, his wife, and their baby), and Nejdet Diş (the husband of her cousin)—you will be forever honored, and one day you'll be avenged. A tribute must also be paid to Jessica Sunshine's mom, Sylvia Sunshine, who was an avid reader, a research librarian by training, and was always surrounded with baskets of books. And a salute, too, for Cristina Thomas's grandmother, Esmira Thomas: she lived for 102 orbits around the Sun, and in that time, she inspired her granddaughter to ask big questions, take risks, have a family, and use telescopes to study the effects of impacting a spacecraft into an asteroid.

So, mum and dad—what is there I can say that will do you both justice? You granted me the sort of upbringing that eludes many other deserving children all over the world. You supported me every which way to make sure that whatever ambitions I had—no matter how madcap—were not fanciful hopes but realizable goals. You aren't just my parents, but my oldest friends, the first to show me the beauty of the world, to show me what it means to belong, to define what a family should be. I have known happiness that defies description, thanks to you. I hope I've made you proud, just as the grandparents were oh so proud of you.

We are all born into families. But, if we are fortunate, we are also chosen by others to start families anew. And for some reason, back in 2014, Stephanie chose me—and in 2024, we got married (in our typically flamboyant fashion) in front of our original families and friends. I don't think I'll ever comprehend what it is that this incredible woman sees in me, a daft rearranger of letters with a pathological inability to take himself seriously. So all I can say to my goofy, whip-smart, tenacious, hilarious, heartwarming, eternal roommate is this: I love you with all my ridiculous heart. Along with our fiercely protective, animated, and exceedingly floofy adopted shollie pup, Lola, you are the family I never knew I needed, and all the proof I require that this world is indeed worth saving.

And, lastly, a message to all those who brazenly stood in our way; to the doubters, the cynics, the little autocrats, the daft and the sociopathic types, the cruel and the petty, the heartless and the thoughtless members of our species: may tiny asteroids fall upon your heads.

NOTES

Prologue: Apocalypse Please

1. Marcus, Robert, et al. "Earth Impact Effects Program." Imperial College London and Purdue University, 2010. https://impact.ese.ic.ac.uk/ImpactEarth/ImpactEffects/.

I. To Russia with Love

1. Kelley, Jay W., et al. "SPACECAST 2020." Defense Technical Information Center (1992). Accessed June 20, 2023.
2. Wikipedia. "The Paradise Syndrome." Wikimedia Foundation. Last modified June 20, 2023. https://en.wikipedia.org/wiki/The_Paradise_Syndrome.
3. Ahrens, Thomas J. "Eugene M. Shoemaker (1928–97)." *Nature*, 1994. Accessed June 20, 2023.
4. "Carolyn Shoemaker (1929–2021)." NASA, August 31, 2021. https://science.nasa.gov/people/carolyn-shoemaker/.
5. Green, Jane. "Impacting Jupiter: The Story of Comet Shoemaker-Levy 9." *BBC Sky at Night Magazine*, May 20, 2019. https://www.skyatnightmagazine.com/space-science/impacting-jupiter-comet-shoemaker-levy-9-25-years-on/.
6. "Carolyn Shoemaker." NASA.
7. "Carolyn Shoemaker." NASA.
8. Green, "Impacting Jupiter."
9. "P/Shoemaker-Levy 9." NASA. July 27, 2021. https://science.nasa.gov/solar-system/comets/p-shoemaker-levy-9/.
10. Caffey, Jim. "1994—The Great Comet Crash of Comet Shoemaker-Levy

9—PBS." March 14, 2017. YouTube Video. https://www.youtube.com/watch?v=YMrWlRRh4rI.

11. Harvey, Ailsa, Tillman, Nola T. "How Big Is Jupiter?" Space.Com, March 23, 2023. https://www.space.com/18392-how-big-is-jupiter.html.
12. Caffey, "1994—The Great Comet Crash."
13. "The Lasting Impacts of Comet Shoemaker-Levy 9." NASA, July 26, 2019. https://science.nasa.gov/science-research/planetary-science/the-lasting-impacts-comet-shoemaker-levy-9/.
14. Browne, Malcolm W. "Earth-Size Storm and Fireballs Shake Jupiter as a Comet Dies." *The New York Times*, July 19, 1994.
15. "P/Shoemaker-Levy 9." NASA.
16. "A Brief History of Comets I (until 1950)." ESO, January 1, 1993. https://www.eso.org/public/unitedkingdom/events/astro-evt/hale-bopp/comet-history-1/.
17. Dobrijevic, Daisy, and Choi, Charles Q. "Comets: Everything You Need to Know about the 'Dirty Snowballs' of Space." Space.com, January 18, 2023. https://www.space.com/comets.html.
18. "Cometary Tails." COSMOS—The SAO Encyclopedia of Astronomy, January 18, 2023. https://astronomy.swin.edu.au/cosmos/c/cometary+tails.
19. "Comets." NASA, December 19, 2019. https://science.nasa.gov/solar-system/comets/facts/.
20. "Arrokoth." NASA, November 13, 2019. https://science.nasa.gov/solar-system/kuiper-belt/arrokoth-2014-mu69/.
21. "Far, Far Away in the Sky: New Horizons Kuiper Belt Flyby Object Officially Named 'Arrokoth'." NASA. November 12, 2019. https://www.nasa.gov/solar-system/far-far-away-in-the-sky-new-horizons-kuiper-belt-flyby-object-officially-named-arrokoth/.
22. "Oort Cloud." NASA, December 15, 2022. https://science.nasa.gov/solar-system/oort-cloud/.
23. *Asteroids III*. Edited by William F. Bottke, Alberto Cellino, Paolo Paolicchi, and Richard P. Binzel. The University of Arizona Press, 2002.
24. "Asteroids." NASA, July 19, 2021. https://science.nasa.gov/solar-system/asteroids/.
25. Tillman, Nola T., and Sutter, Paul. "Asteroid Belt: Facts & Formation." Space.com. May 5, 2017. https://www.space.com/16105-asteroid-belt.html.
26. Andrews, Robin G. "This Is What It Looks Like When an Asteroid Gets Destroyed." *The New York Times*, November 26, 2019.
27. Joel, Lucas. "The Dinosaur-Killing Asteroid Acidified the Ocean in a Flash." *The New York Times*, October 21, 2019.
28. Brannen, Peter. *The Ends of the World*. HarperCollins, 2018.
29. Raghavan, Akila. "Giant Tsunami from Dino-Killing Asteroid Impact Revealed in Fossilized 'Megaripples'." *Science*, July 12, 2021. https://www.science.org/

content/article/giant-tsunami-dino-killing-asteroid-impact-revealed-fossilized-megaripples.

30. Greshko, Michael. "Dinosaur-Killing Asteroid Most Likely Struck in Spring." *National Geographic*, February 23, 2022. https://www.nationalgeographic.com/science/article/dinosaur-killing-asteroid-most-likely-struck-in-spring?loggedin=true.
31. Alvarez, Luis W., et al. "Extraterrestrial Cause for the Cretaceous-Tertiary Extinction." *Science*, 1980. Accessed June 20, 2023. https://www.science.org/doi/10.1126/science.208.4448.1095.
32. Hildebrand, Alan R. "Chicxulub Crater: A Possible Cretaceous/Tertiary Boundary Impact Crater on the Yucatán Peninsula, Mexico." *Geology*, 1994. Accessed June 20, 2023. https://pubs.geoscienceworld.org/gsa/geology/article-abstract/19/9/867/205322/Chicxulub-Crater-A-possible-Cretaceous-Tertiary.
33. Boslough, Mark, and Kring, David A. "Chelyabinsk: Portrait of an Asteroid Airburst." *Physics Today*, 2014. Accessed June 20, 2023. https://pubs.aip.org/physicstoday/article/67/9/32/414872/Chelyabinsk-Portrait-of-an-asteroid-airburstVideo.
34. Alves, Artur. "Russian Meteor 15-02-2013 (Best Shots) [HD]." YouTube Video, February 15, 2013. https://www.youtube.com/watch?v=fBLjB5qavxY.
35. Popova, Olga P., et al. "Chelyabinsk Airburst, Damage Assessment, Meteorite Recovery, and Characterization." *Science*, 2013. Accessed June 20, 2023. https://www.science.org/doi/10.1126/science.1242642.
36. Peplow, Mark. "Military History: Dinner at the Fission Chips." *Nature*, 2013. Accessed June 20, 2023. https://www.nature.com/articles/495444a.
37. Popova, "Chelyabinsk Airburst."
38. "Additional Details on the Large Feb. 15 Fireball over Russia." Jet Propulsion Laboratory, February 15, 2013. https://www.jpl.nasa.gov/news/additional-details-on-the-large-feb-15-fireball-over-russia.
39. Jenniskens, Peter, et al. "Tunguska Eyewitness Accounts, Injuries, and Casualties." *Icarus*, 2019. Accessed June 20, 2023. https://www.sciencedirect.com/science/article/abs/pii/S0019103518305104.
40. Weisberger, Mindy. "Meteor That Blasted Millions of Trees in Siberia Only 'Grazed' Earth, New Research Says." *LiveScience*, May 6, 2020. https://www.livescience.com/tunguska-impact-explained.html.
41. Kelley. "SPACECAST 2020."
42. "NEO (Near-Earth Object." Center for Near Earth Object Studies (CNEOS). https://cneos.jpl.nasa.gov/glossary/NEO.html.
43. "Near-Earth Asteroids as of August 2023." NASA, August 31, 2023. https://www.nasa.gov/directorates/smd/planetary-science-division/planetary-defense-coordination-office/near-earth-asteroids-as-of-august-31-2023/.
44. Prabhakar, Arati, et al. "National Preparedness Strategy for Near-Earth Object

Hazards and Planetary Defense." National Science & Technology Council, 2023. Accessed June 20, 2023.

45. Prabhakar, "National Preparedness Strategy."
46. Prabhakar, "National Preparedness Strategy."
47. Smith, Marcia. "NASA'S New NEO Mission Will Substantially Reduce Time to Find Hazardous Asteroids." SpacePolicyOnline, January 19, 2020.
48. "Planetary Defense at NASA." NASA. April 1, 2019. https://science.nasa.gov/planetary-defense/.

II. This Is for the Dinosaurs

1. "The Story so Far." European Space Agency (ESA), June 20, 2023. https://www.esa.int/Space_Safety/Hera/The_story_so_far.
2. O'Callaghan, Jonathan. "Behold the Weird Physics of Double-Impact Asteroids." *Wired*, June 27, 2022. https://www.wired.com/story/mars-binary-asteroid-craters/.
3. Pravec, Petr, et al. "Photometric Observations of the Binary Near-Earth Asteroid (65803) Didymos in 2015–2021 Prior to DART Impact." *The Planetary Science Journal*, 2022. Accessed June 20, 2023. https://iopscience.iop.org/article/10.3847/PSJ/ac7be1.
4. "DART." Johns Hopkins University Applied Physics Laboratory (JHUAPL), June 20, 2023. https://dart.jhuapl.edu/Mission/index.php.
5. "Double Asteroid Redirection Test." NASA, October 28, 2022. https://nssdc.gsfc.nasa.gov/nmc/spacecraft/display.action?id=2021-110A.
6. "SMART Nav." JHUAPL, October 1, 2019. https://doi.org/https://www.jhuapl.edu/interactive/navigating-double-asteroid-redirection-test-on-its-own.
7. Amanda S. Lecture Screenshot. Twitter, August 19, 2022. https://twitter.com/acstadermann/status/1560613013555535874.
8. Dreier, Casey. "How Much Does the James Webb Space Telescope Cost?" The Planetary Society, October 25, 2021. https://www.planetary.org/articles/cost-of-the-jwst.
9. Foust, Jeff. "NASA Presses Ahead with Asteroid Mission despite ESA Funding Decision." *SpaceNews*, December 13, 2016. https://spacenews.com/nasa-presses-ahead-with-asteroid-mission-despite-esa-funding-decision/.
10. "Hera Mission." Hera Mission, January 1, 2021. https://www.heramission.space/.
11. "What Are SmallSats and CubeSats?" NASA, February 26, 2016. https://www.nasa.gov/content/what-are-smallsats-and-cubesats/.
12. "LICIACube." Agenzia Spaziale Italiana (ASI), January 1, 2019. https://www.asi.it/en/planets-stars-universe/solar-system-and-beyond/liciacube/.
13. Allen, Jennifer. "Crash at Crush Killed 3 in 1896 Texas Publicity Stunt." *Trains*, September 10, 2021.

14. Andrews, Robin. "NASA'S DART Mission Could Help Cancel an Asteroid Apocalypse." *Scientific American*, December 11, 2021. https://www.scientificamerican.com/article/nasas-dart-mission-could-help-cancel-an-asteroid-apocalypse/.
15. Andrews, "NASA'S DART Mission."
16. Rivkin, Andrew. "Asteroids Have Been Hitting the Earth for Billions of Years. In 2022, We Hit Back." The Planetary Society, September 27, 2018. https://www.planetary.org/articles/hitting-back-dart.
17. "DART, NASA's Test to Stop an Asteroid from Hitting Earth." The Planetary Society, October 1, 2022. https://www.planetary.org/space-missions/dart.
18. "Countries with the Highest Military Spending Worldwide in 2022." Statista. June 5, 2023. https://www.statista.com/statistics/262742/countries-with-the-highest-military-spending/.

III. Hollywood and the Nuclear Hail Mary

1. Horan, Lansing S. IV, et al. "Impact of Neutron Energy on Asteroid Deflection Performance." *Acta Astronautica*, 2021. Accessed June 20, 2023. https://scholar.afit.edu/cgi/viewcontent.cgi?article=1738&context=facpub.
2. "Top 10 Things You Didn't Know about LLNL." Energy.gov. https://www.energy.gov/articles/top-10-things-you-didnt-know-about-lawrence-livermore-national-laboratory.
3. "About LLNL." Lawrence Livermore National Laboratory (LLNL). https://www.llnl.gov/about.
4. "How NIF Works." LLNL. https://lasers.llnl.gov/about/how-nif-works.
5. "Top 10 Things." Energy.gov.
6. "Trinity: World's First Nuclear Test." AFNWC. https://www.afnwc.af.mil/About-Us/History/Trinity-Nuclear-Test/.
7. "Trinity Test Eyewitnesses." Atomic Heritage Foundation. https://ahf.nuclearmuseum.org/ahf/key-documents/trinity-test-eyewitnesses/.
8. Horgan, John. "Bethe, Teller, Trinity and the End of Earth." *Scientific American*, August 4, 2015. https://blogs.scientificamerican.com/cross-check/bethe-teller-trinity-and-the-end-of-earth/.
9. "The Outer Space Treaty." United Nations Office for Outer Space Affairs (UNOOSA). https://www.unoosa.org/oosa/en/ourwork/spacelaw/treaties/introouterspacetreaty.html.
10. Glasstone, S., and P. J. Nolan. "The Effects of Nuclear Weapons. Third Edition" U.S. Department of Energy, Office of Scientific and Technical Information (OSTI), 1977. Accessed June 20, 2023.
11. Andrews, Robin G. "How a Nuclear Bomb Could Save Earth From a Stealthy Asteroid." *The New York Times*, October 18, 2021.

12. Guiterrez, Brian. "Why the U.S. Once Set off a Nuclear Bomb in Space." *National Geographic*, July 15, 2021.
13. Andrews, Robin G. "If We Blow Up an Asteroid, It Might Put Itself Back Together." *The New York Times*, March 8, 2019.
14. Horan, Lansing S. IV, et al. "Neutron Energy Effects on Asteroid Deflection." Lecture, April 1, 2021.
15. "The Great Escape: SLS Provides Power for Missions to the Moon." NASA. https://www.nasa.gov/humans-in-space/space-launch-system/the-great-escape-sls-provides-power-for-missions-to-the-moon-duzxi/#:~:text=SLS%20uses%20its%20power%20to,around%20eight%20minutes%20after%20launch.
16. Temperton, James. "'Now I Am Become Death, the Destroyer of Worlds.' The Story of Oppenheimer's Infamous Quote." *Wired*, August 9, 2017. https://www.wired.co.uk/article/manhattan-project-robert-oppenheimer.
17. Temperton, "Now I Am Become Death."
18. Bruck Syal, Megan. Twitter, April 5, 2023. https://twitter.com/meganimpact/status/1643628408969089025.
19. "With Its Single 'Eye,' NASA's DART Returns First Images from Space." NASA, February 12, 2021. https://www.nasa.gov/feature/with-its-single-eye-nasa-s-dart-returns-first-images-from-space/.

IV. Spying on Heaven

1. "How Do Telescopes Work?" Museums Victoria—ScienceWorks. https://museumsvictoria.com.au/scienceworks/visiting/melbourne-planetarium/fact-sheets/how-do-telescopes-work.
2. "NEO Observations Program." Planetary Defense Coordination Office (PDCO). NASA. https://science.nasa.gov/planetarydefense/neoo.
3. "Pan-STARRS." Institute for Astronomy, University of Hawaii. https://www2.ifa.hawaii.edu/research/Pan-STARRS.shtml.
4. "The ATLAS Project." Asteroid Terrestrial-impact Last Alert System (ATLAS). https://atlas.fallingstar.com/home.php.
5. "Discovery Statistics." Center for Near Earth Object Studies (CNEOS). April 20, 2023. https://cneos.jpl.nasa.gov/stats/site_all.html.
6. "1p/Halley." NASA, September 16, 2022. https://solarsystem.nasa.gov/asteroids-comets-and-meteors/comets/1p-halley/in-depth/.
7. "NASA Program Predicted Impact of Small Asteroid Over Ontario, Canada." NASA. November 22, 2022. https://www.nasa.gov/solar-system/asteroids/nasa-program-predicted-impact-of-small-asteroid-over-ontario-canada/.
8. "Overview of the Verification Regime." Comprehensive Nuclear-Test-Ban Treaty Organization (CTBTO). https://www.ctbto.org/our-work/verification-regime.

9. "Fireballs." CNEOS, May 20, 2023. https://cneos.jpl.nasa.gov/fireballs/.
10. "Sentry." CNEOS. https://cneos.jpl.nasa.gov/sentry/.
11. "About Asteroids and Planetary Defence." European Space Agency (ESA). https://www.esa.int/Space_Safety/About_asteroids_and_Planetary_Defence.
12. Perozzi, E., B. Borgia, and M. Micheli. "The European NEO Coordination Centre." Memorie Della Societa Astronomica Italiana, 2016. Accessed June 20, 2023.
13. "Flyeye: ESA's Bug-eyed Asteroid Hunter." ESA. https://www.esa.int/Space_Safety/Flyeye_ESA_s_bug-eyed_asteroid_hunter.
14. "NEOCC Orbit Determination and Impact Monitoring Software Update." Near-Earth Objects Coordination Centre (NEOCC), December 20, 2022. https://neo.ssa.esa.int/-/neocc-orbit-determination-and-impact-monitoring-software-update.
15. "Predicted Impact Point and Time, Computed by ESA's Imminent Impactor Alert System 'Meerkat'." ESA, March 15, 2022. https://www.esa.int/ESA_Multimedia/Images/2022/03/Predicted_impact_point_and_time_computed_by_ESA_s_imminent_impactor_alert_system_Meerkat.
16. "Risk List." NEOCC, April 30, 2022. https://doi.org/https://neo.ssa.esa.int/risk-list.
17. EAPS. "Richard Binzel: Eyes on the Skies and a Passion for Planetary Science." MIT News, February 22, 2022. https://news.mit.edu/2022/richard-binzel-eyes-on-skies-passion-for-planetary-science-0222.
18. "Minor Planet Names." Minor Planet Center, May 10, 2023. https://minorplanetcenter.net/iau/lists/MPNames.html.
19. "Torino Impact Hazard Scale." CNEOS. https://cneos.jpl.nasa.gov/sentry/torino_scale.html.
20. Reich, Aaron. "Asteroid the Size of a Giraffe to Skim Past Earth This Week." *The Jerusalem Post*, June 1, 2022.
21. Reich, Aaron. "Corgi-Sized Meteor as Heavy as 4 Baby Elephants Hit Texas." *The Jerusalem Post*, February 21, 2023.
22. Reich, Aaron. "2 Asteroids the Size of 100 Pugs to Pass Earth Tuesday." *The Jerusalem Post*, January 22, 2023.
23. Andrews, Robin G. "A 22-Million-Year Journey From the Asteroid Belt to Botswana." *The New York Times*, April 29, 2021.
24. Jenniskens, Peter. "The Impact and Recovery of Asteroid 2008 TC3." *Nature*, 2009. Accessed June 20, 2023. https://www.nature.com/articles/nature07920.
25. Andrews, "A 44-Million-Year Journey."
26. "Central Kalahari Game Reserve." Botswana Tourism. https://www.botswanatourism.co.bw/explore/central-kalahari-game-reserve.
27. Andrews, "A 44-Million-Year Journey."
28. Andrews, "A 44-Million-Year Journey."
29. "Meteors and Meteorites." NASA, December 19, 2019. https://solarsystem.nasa.gov/asteroids-comets-and-meteors/meteors-and-meteorites/in-depth/.

30. Andrews, Robin G. "The Mystery of Antarctica's Missing Meteorites." *The Atlantic,* February 26, 2019. https://www.theatlantic.com/science/archive/2019/02/hunting-for-antarcticas-lost-meteorites/583564/.
31. Russell, Sara S., et al. "Recovery and Curation of the Winchcombe (CM2) Meteorite." *Meteorics & Planetary Science,* 2023. Accessed June 20, 2023. https://onlinelibrary.wiley.com/doi/10.1111/maps.13956#.Y_3AJzqs5Aw.twitter.
32. "The UK Fireball Alliance." UKFall. https://www.ukfall.org.uk/.
33. Russell, "Recovery and Curation."
34. Russell, "Recovery and Curation."
35. Andrews, Robin G. "Newfound Meteorite Could Help Unlock Secrets of the Solar System." *National Geographic,* March 11, 2021. https://www.nationalgeographic.com/science/article/newfound-meteorite-could-help-unlock-secrets-of-the-solar-system.
36. Amos, Jonathan. "A Fireball, a Driveway and a Priceless Meteorite." *BBC News,* March 9, 2021. https://www.bbc.co.uk/news/science-environment-56337876.
37. Andrews, "Newfound Meteorite."
38. Russell, "Recovery and Curation."
39. King, Ashley J., et al. "The Winchcombe Meteorite, a Unique and Pristine Witness from the Outer Solar System." *Science Advances,* 2022. Accessed June 20, 2023. https://www.science.org/doi/10.1126/sciadv.abq3925.
40. Amos, Jonathan. "Winchcombe Meteorite Fragments Auctioned for More than Gold." *BBC News,* February 23, 2022. https://www.bbc.co.uk/news/science-environment-60495954.
41. Amos, Jonathan. "Winchcombe Meteorite Driveway to Go on Display." *BBC News,* September 8, 2021. https://www.bbc.co.uk/news/science-environment-58493430.

V. Never Tell Me the Odds

1. John, Jeremy, et al. "NEXT-C Lessons Learned on the DART Mission for Future Integration and Test." IEEE, 2023. Accessed June 20, 2023.
2. "9P/Tempel 1." NASA. December 19, 2021. https://solarsystem.nasa.gov/asteroids-comets-and-meteors/comets/9p-tempel-1/in-depth/.
3. "Jupiter-Family Comets." COSMOS—The SAO Encyclopedia of Astronomy. https://astronomy.swin.edu.au/cosmos/J/Jupiter-family+comets.
4. Leary, Warren E. "Spacecraft Is on a Collision Course With a Comet, Intentionally." *The New York Times,* June 28, 2005.
5. "Deep Impact (EPOXI)." NASA. https://solarsystem.nasa.gov/missions/deep-impact-epoxi/in-depth/.
6. Lakdawalla, Emily. "Deep Impact On Course for Comet Crash; Mission Is Already Producing Science Returns." The Planetary Society, July 1, 2005. https://www.planetary.org/articles/0121.

7. "67P/Churyumov-Gerasimenko." NASA. August 8, 2022. https://solarsystem.nasa.gov/asteroids-comets-and-meteors/comets/67p-churyumov-gerasimenko/in-depth/.
8. "67P." NASA.
9. "Rosetta-Philae." NASA. https://solarsystem.nasa.gov/missions/rosetta-philae/in-depth/.
10. Jia, Pan, Bruno Andreotti, and Philippe Claudin. "Giant Ripples on Comet 67P/Churyumov–Gerasimenko Sculpted by Sunset Thermal Wind." *PNAS*, 2016. Accessed June 20, 2023. https://www.pnas.org/doi/10.1073/pnas.1612176114.
11. "Rosetta-Philae." NASA.
12. "Philae Obelisk." Wikimedia Foundation. Last modified March 5, 2023. https://en.wikipedia.org/wiki/Philae_obelisk.
13. "Landing on a Comet." European Space Agency (ESA). September 1, 2019. https://sci.esa.int/web/rosetta/-/54470-landing-on-a-comet.
14. "Rosetta's Comet Lander Landed Three Times." JPL. November 13, 2014. https://www.jpl.nasa.gov/news/rosettas-comet-lander-landed-three-times.
15. "67P." NASA.
16. Amos, Jonathan. "Winchcombe Meteorite Bolsters Earth Water Theory." *BBC News*, November 16, 2022. https://www.bbc.co.uk/news/science-environment-63631563.
17. Amos, Jonathan. "Philae: Lost Comet Lander Is Found." *BBC News*, September 6, 2016. https://www.bbc.com/news/science-environment-37276221.
18. Gibney, Elizabeth. "Mission Accomplished: Rosetta Crashes into Comet." *Nature*, September 30, 2016. https://www.nature.com/articles/nature.2016.20705.
19. O'Rourke, Laurence, et al. "The Philae Lander Reveals Low-Strength Primitive Ice inside Cometary Boulders." *Nature*, (2020). Accessed June 20, 2023. https://www.nature.com/articles/s41586-020-2834-3.
20. Gibney, Elizabeth. "'Like Froth on a Cappuccino': Spacecraft's Chaotic Landing Reveals Comet's Softness." *Scientific American*, November 1, 2020. https://www.scientificamerican.com/article/like-froth-on-a-cappuccino-spacecrafts-chaotic-landing-reveals-comets-softness/.
21. Andrews, Robin. "NASA'S DART Mission Could Help Cancel an Asteroid Apocalypse." *Scientific American*, December 11, 2021. https://www.scientificamerican.com/article/nasas-dart-mission-could-help-cancel-an-asteroid-apocalypse/.
22. Mann, Adam. "Asteroid Ryugu: The Twirling Space Rock Visited by Hayabusa2." *Space.com*, January 12, 2020. https://www.space.com/asteroid-ryugu.
23. "C-type Asteroids Facts & Information." The Nine Planets. March 5, 2020. https://nineplanets.org/c-type-asteroids/.
24. Plait, Phil. "Ryuga, Front to Back. Syfy, July 14, 2018. https://www.syfy.com/syfy-wire/ryugu-front-to-back.
25. "Hayabusa2." NASA. December 8, 2021. https://solarsystem.nasa.gov/missions/hayabusa-2/in-depth/.

26. "Hayabusa2." NASA.
27. Rayne, Elizabeth. "Yes Japan Just Bombed an Asteroid." Syfy, April 5, 2019. https://www.syfy.com/syfy-wire/japan-just-bombed-an-asteroid.
28. "Hayabusa2 Impact Crater Event." SpacePolicyOnline.com, April 3, 2019. https://spacepolicyonline.com/events/hayabusa2-impact-crater-event-apr-4-2019-asteroid-ryugu-1036-pm-et/.
29. Bartels, Meghan. "Boop! Japanese Spacecraft Grabs Second Sample from Asteroid Ryugu." Space.com, July 11, 2019. https://www.space.com/japan-hayabusa2-asteroid-sample-complete-july-2019.html.
30. Chang, Kenneth. "Japan's Journey to an Asteroid Ends With a Hunt in Australia's Outback." *The New York Times*, December 5, 2020.
31. "Hayabusa." NASA. January 25, 2018. https://solarsystem.nasa.gov/missions/hayabusa/in-depth/.
32. "Japan's Postwar Constitution." Council for Foreign Relations (CFR). January 25, 2018. https://www.cfr.org/japan-constitution/japans-postwar-constitution/.
33. Oi, Mariko. "Japan's Contradictory Military Might." *BBC News*, March 15, 2012. https://www.bbc.co.uk/news/world-asia-17175834.
34. "Japan Passes Law Allowing Troops to Fight Abroad." *Al Jazeera*, September 19, 2015. https://www.aljazeera.com/news/2015/9/19/japan-passes-law-allowing-troops-to-fight-abroad.
35. "Ten Things to Know About Bennu." NASA. October 16, 2020. https://www.nasa.gov/feature/goddard/2020/bennu-top-ten/.
36. "OSIRIS-REx." NASA, July 7, 2022. https://solarsystem.nasa.gov/missions/osiris-rex/in-depth/.
37. "OSIRIS-REx, NASA's Sample Return Mission to Asteroid Bennu." The Planetary Society, September 1, 2021. https://www.planetary.org/space-missions/osiris-rex.
38. "NASA'S OSIRIS-REx Begins Its Countdown to TAG." NASA, September 24, 2020. https://www.nasa.gov/feature/goddard/2020/osiris-rex-begins-its-countdown-to-tag.
39. Chang, Kenneth. "Seeking Solar System's Secrets, NASA'S OSIRIS-REX Mission Touches Bennu Asteroid." *The New York Times*, October 20, 2020.
40. "X Marks the Spot: NASA Selects Site for Asteroid Sample Collection." NASA, December 12, 2019. https://www.nasa.gov/press-release/x-marks-the-spot-nasa-selects-site-for-asteroid-sample-collection/.
41. Wolner, CWV. "A Time Capsule from the Early Solar System Is En Route to Earth." *Eos*, August 4, 2022. https://eos.org/science-updates/a-time-capsule-from-the-early-solar-system-is-en-route-to-earth.
42. "Surprise—Again! Asteroid Bennu Reveals Its Surface Is Like a Plastic Ball Pit." NASA, July 7, 2022. https://www.nasa.gov/feature/goddard/2022/surprise-again-asteroid-bennu-reveals-its-surface-is-like-a-plastic-ball-pit/.

43. Wolner, "A Time Capsule."
44. Chang, Kenneth. "NASA Says an Asteroid Will Have a Close Brush With Earth. But Not Until the 2100s." *The New York Times,* August 11, 2021.
45. "101955 Bennu (1999 RQ36)—Earth Impact Risk Summary." CNEOS, June 21, 2023. https://cneos.jpl.nasa.gov/sentry/details.html#?des=101955.
46. "Asteroid Apophis: Will It Hit Earth? Your Questions Answered." The Planetary Society. https://www.planetary.org/articles/will-apophis-hit-earth.
47. Jha, Alok. "It's Called Apophis. It's 390m Wide. And It Could Hit Earth in 31 Years' Time." *The Guardian,* December 7, 2005.
48. "NASA Analysis: Earth Is Safe From Asteroid Apophis for 100-Plus Years." NASA, March 26, 2021. https://www.nasa.gov/solar-system/nasa-analysis-earth-is-safe-from-asteroid-apophis-for-100-plus-years/.

VI. This Is How We Lose

1. Manni, Gregory. "Ludovic Ferrière, Vienna's Interplanetary Explorer." *Metropole,* October 12, 2021. https://metropole.at/ludovic-ferriere-viennas-interplanetary-explorer/.
2. Xu, Jessica. "The Chicxulub Impact: Impact Energy and Climate Change." Stanford University, July 12, 2015. http://large.stanford.edu/courses/2015/ph240/xu2/.
3. Amos, Jonathan. "Chicxulub 'dinosaur' Crater Drill Project Declared a Success." *BBC News,* May 25, 2016. https://www.bbc.co.uk/news/science-environment-36377679.
4. Ferrière, Ludovic, et al. "The Newly Confirmed Luizi Impact Structure, Democratic Republic of Congo—Insights into Central Uplift Formation and Post-impact Erosion." *Geology,* 2011. Accessed June 21, 2023. https://pubs.geoscienceworld.org/gsa/geology/article-abstract/39/9/851/130676/The-newly-confirmed-Luizi-impact-structure?redirectedFrom=fulltext.
5. "Democratic Republic of Congo. Overview." World Bank, March 29, 2023. https://www.worldbank.org/en/country/drc/overview.
6. "Democratic Republic of Congo." Gov.UK, Foreign Travel Advice, January 18, 2024. https://www.gov.uk/foreign-travel-advice/democratic-republic-of-the-congo.
7. Ferrière, "Luizi Impact Structure."
8. Andrews, Robin G. "Mount Nyiragongo Just Erupted—Here's Why It's One of Africa's Most Dangerous Volcanoes." *National Geographic,* May 24, 2021. https://www.nationalgeographic.com/science/article/mount-nyiragongo-just-erupted-why-its-one-of-africas-most-dangerous-volcanoes?loggedin=true&rnd=1684344510805.
9. McKie, Robin. "Poisonous Gas from African Lake Poses Threat to Millions." *The Observer,* July 26, 2009. https://www.theguardian.com/world/2009/jul/26/africa-lake-kivu-co2-gas.

10. "Democratic Republic of Congo." Gov.UK, Foreign Travel Advice.
11. "Gabon." Gov.UK, Foreign Travel Advice, April 28, 2023. https://www.gov.uk/foreign-travel-advice/gabon.
12. Andrews, Robin G. "The Largest Comet Ever Found Is Making Its Move Into a Sky Near You." *The New York Times*, July 28, 2021.
13. Andrews, Robin G. "Hubble Telescope Zooms In on the Biggest Comet Ever Spotted." *The New York Times*, April 14, 2022.
14. Artemieva, Natalia, and Elisabetta Pierazzo. "The Canyon Diablo Impact Event: Projectile Motion through the Atmosphere." *Meteorics & Planetary Science*, (2010). Accessed June 21, 2023. https://onlinelibrary.wiley.com/doi/10.1111/j.1945-5100.2009.tb00715.x.
15. Barringer, Brandon. "Daniel Moreau Barringer (1860–1929) and His Crater." *Meteorics Science*, 1964. Accessed June 21, 2023. https://articles.adsabs.harvard.edu//full/1964Metic . . . 2..183B/0000183.000.html.
16. "Meteor Crater." American Museum of Natural History (AMNH). https://www.amnh.org/exhibitions/permanent/meteorites/meteorite-impacts/meteor-crater.
17. Chao, E. C. T., E. M. Shoemaker, and B. M. Madsen. "First Natural Occurrence of Coesite." *Science*, 1960. Accessed June 21, 2023. https://www.science.org/doi/abs/10.1126/science.132.3421.220.
18. "DART Team Confirms Orbit of Targeted Asteroid." NASA, August 5, 2022. https://www.nasa.gov/feature/dart-team-confirms-orbit-of-targeted-asteroid/.
19. "DART Tests Autonomous Navigation System Using Jupiter and Europa." NASA, September 22, 2022. https://www.nasa.gov/feature/dart-tests-autonomous-navigation-system-using-jupiter-and-europa/.

VII. A Dress Rehearsal for Saving the World

1. Overbye, Dennis. "How Do You Tell the World That Doomsday Has Arrived?" *The New York Times*, December 9, 2021.
2. Overbye, "How Do You Tell the World."
3. Browne, Malcolm W. "Asteroid Is Expected to Make A Pass Close to Earth in 2028." *The New York Times*, March 12, 1998.
4. Overbye, "How Do You Tell the World."
5. "2022 Interagency Tabletop Exercise (PD TTX4)." Center for Near Earth Object Studies CNEOS. March 1, 2022. https://cneos.jpl.nasa.gov/pd/cs/ttx22/.
6. "'Ask Not What Your Country can Do for Your . . .'" John F. Kennedy Presidential Library and Museum. https://www.jfklibrary.org/learn/education/teachers/curricular-resources/ask-not-what-your-country-can-do-for-you.
7. "We Are FEMA: Leviticus 'L.A.' Lewis." FEMA, November 10, 2020. https://www.fema.gov/blog/we-are-fema-leviticus-la-lewis.
8. Wikipedia. "Winston-Salem, North Carolina." Wikimedia Foundation.

Last modified December 1, 2023. https://en.wikipedia.org/wiki/Winston-Salem,_North_Carolina.

9. "2022 Interagency Tabletop Exercise." CNEOS.
10. "2022 Interagency Tabletop Exercise." CNEOS.
11. "NASA Policy Directive, NPD 8740.1." NASA, NODIS Library, February 15, 2022. https://nodis3.gsfc.nasa.gov/displayDir.cfm?t=NPD&c=8740&s=1#:~:text=It%20is%20NASA%20policy%20to,Science%20and%20Technology%20Policy%20(OSTP).
12. "IAWN: About." International Asteroid Warning Network (IAWN). https://iawn.net/about.shtml.
13. "Summary of the 7th Meeting of the Space Mission Planning Advisory Group." European Space Agency (ESA). October 14, 2016. https://www.cosmos.esa.int/web/smpag/meeting_07_-_oct_2016.
14. "2022 Interagency Tabletop Exercise." CNEOS.
15. "Countering Truth Decay." RAND. https://www.rand.org/research/projects/truth-decay.html.
16. Vale, Paul. "John Oliver Nails Climate Change Deniers On HBO Show." *The Huffington Post*, May 12, 2014. https://www.huffingtonpost.co.uk/2014/05/12/john-oliver-climate-change-deniers_n_5311401.html.
17. McArdle, Megan. "We Finally Know for Sure that Lies Spread Faster than the Truth. This Might Be Why." *The Washington Post*, March 14, 2018.
18. "2015 IAA Planetary Defense Conference." IAA. May 1, 2015. https://iaaspace.org/wp-content/uploads/iaa/Scientific%20Activity/pdcreportfinal.pdf.
19. "2015 Planetary Defense Conference." IAA.
20. "2022 Interagency Tabletop Exercise." CNEOS.
21. "2022 Interagency Tabletop Exercise." CNEOS.
22. Andrews, Robin G. "Pets Are Like Family, So Why Do They Get Left Behind During Disasters?" *Gizmodo*, May 15, 2019. https://gizmodo.com/pets-are-like-family-so-why-do-they-get-left-behind-du-1834715712.
23. Marcus, Robert, H. Jay Melosh, and Gareth Collins. "Earth Impact Effects Program." Imperial College London and Purdue University, 2010. https://impact.ese.ic.ac.uk/ImpactEarth/ImpactEffects/.
24. "2022 Interagency Tabletop Exercise." CNEOS.
25. "Planetary Defense Conference Exercise—2021." CNEOS, April 26, 2021. https://cneos.jpl.nasa.gov/pd/cs/pdc21/.

VIII. In the Shadow of the Sun

1. "Leonardo DiCaprio—Rage & Anger Scene From Movie—Don't Look Up." Movieza, December 25, 2021. YouTube Video, https://www.youtube.com/watch?v=HonX5tKui9Q.

2. "WISE—Mission." WISE, December 1, 2010. http://wise.ssl.berkeley.edu/mission.html.
3. "Why Infrared?" University of Arizona, NEO Surveyor. https://neos.arizona.edu/mission/why-infrared.
4. Andrews, Robin G. "'Planet Killer' Asteroid Spotted That Poses Distant Risk to Earth." *The New York Times*, October 31, 2022.
5. Andrews, Robin G. "Burying CAESAR." *Scientific American*, July 25, 2019. https://www.scientificamerican.com/article/burying-caesar-how-nasa-picks-winners-and-losers-in-space-exploration/.
6. Williams, Matt. "The Next Generation of Exploration: The NEOCam Mission." Universe Today, October 9, 2015. https://www.universetoday.com/122785/the-next-generation-of-exploration-the-neocam-mission/.
7. Smith, Marcia. "NASA Announces New Mission to Search for Asteroids." *SpacePolicyOnline*, September 23, 2019. https://spacepolicyonline.com/news/nasa-announces-new-mission-to-search-for-asteroids/.
8. Foust, Jeff. "NASA to Develop Mission to Search for Near-Earth Asteroids." *SpaceNews*, September 23, 2019. https://spacenews.com/nasa-to-develop-mission-to-search-for-near-earth-asteroids/.
9. Smith, "NASA Announces New Mission."
10. "Dr. Amy Mainzer." NASA. https://science.nasa.gov/science-committee/members/dr-amy-mainzer. Last accessed in July 2023.
11. "SEH 3.0 NASA Program/Project Life Cycle." NASA, December 12, 2019. https://www.nasa.gov/seh/3-project-life-cycle/.
12. Dreier, Casey. "NASA Abruptly Delays a Critical Planetary Defense Mission." The Planetary Society, December 17, 2020. https://www.planetary.org/articles/nasa-abruptly-delays-neosm.
13. Foust, Jeff. "NASA Asteroid Hunter Mission Moves into Next Phase of Development." *SpaceNews*, July 16, 2021. https://spacenews.com/nasa-asteroid-hunter-mission-moves-into-next-phase-of-development/.
14. "NASA's NEO Surveyor Successfully Passes Key Milestone." NASA, December 6, 2022. https://blogs.nasa.gov/neosurveyor/2022/12/06/nasas-neo-surveyor-successfully-passes-key-milestone/.
15. "SHE 3.0 NASA Program/Project." NASA.
16. "NASA's NEO Surveyor." NASA.
17. Fitzpatrick, Jack. "White House Wants NASA to Slow Hunt for Killer Asteroids in 'Baffling' Move." *Bloomberg*, September 1, 2022. https://www.bloomberg.com/news/articles/2022-09-01/killer-asteroid-search-hits-bump-as-biden-team-urges-nasa-delay?leadSource=uverify%20wall.
18. Hall, Shannon. "NASA's Interplanetary Plans May Be Lurching toward Disaster." *Scientific American*, May 9, 2023. https://www.scientificamerican.com/article/nasas-interplanetary-plans-may-be-lurching-toward-disaster/.

19. "What Happened with Psyche?" The Planetary Society, May 4, 2023. https://www.planetary.org/articles/what-happened-with-psyche#:~:text=Psyche%2C%20a%20mission%20to%20an,Venus%20which%20was%20delayed%20indefinitely.
20. Dreier, Casey. "NEO Surveyor Is Confirmed." The Planetary Society, December 14, 2022. https://www.planetary.org/articles/neo-surveyor-is-confirmed.
21. Dreier, Casey. "How NASA's Planetary Defense Budget Grew By More Than 4000% in 10 Years." The Planetary Society, September 26, 2019. https://www.planetary.org/articles/nasas-planetary-defense-budget-growth.
22. Cowing, Keith. "NASA's Search for Asteroids to Help Protect Earth and Understand Our History." SpaceRef, April 22, 2014. https://spaceref.com/science-and-exploration/nasas-search-for-asteroids-to-help-protect-earth-and-understand-our-history/.
23. Dreier, "NASA's Planetary Defense Budget."
24. Dreier, "NASA's Planetary Defense Budget."
25. "Countries with the Highest Military Spending Worldwide in 2022." Statista. https://www.statista.com/statistics/262742/countries-with-the-highest-military-spending/.
26. Johnson, Courtney. "How Americans See the Future of Space Exploration, 50 Years after the First Moon Landing." Pew Research Center, July 17, 2019. https://www.pewresearch.org/short-reads/2019/07/17/how-americans-see-the-future-of-space-exploration-50-years-after-the-first-moon-landing/.
27. Binzel, Richard, Donald Yeomans, and Timothy Swindle. "Op-ed | A Space-Based Survey, Not Luck, Must Be Our Plan against Hazardous Asteroids." *SpaceNews*, October 12, 2018. https://spacenews.com/op-ed-a-space-based-survey-not-luck-must-be-our-plan-against-hazardous-asteroids/.
28. Overbye, Dennis. "Vera Rubin, 88, Dies; Opened Doors in Astronomy, and for Women." *The New York Times*, December 27, 2016.
29. "About Rubin Observatory." Vera C. Rubin Observatory. https://www.lsst.org/about.
30. "Ceres." NASA. https://science.nasa.gov/mission/dawn/science/ceres/.
31. "The Launch of Sputnik, 1957." U.S. Department of State. https://2001-2009.state.gov/r/pa/ho/time/lw/103729.htm#:~:text=On%20October%204%2C%201957%2C%20the,first%20artificial%20satellite%2C%20Sputnik%20I.
32. Howell, Elizabeth, and Tereza Pultarova. "Starlink Satellites: Everything You Need to Know about the Controversial Internet Megaconstellation." *Space.com*, July 25, 2023. https://www.space.com/spacex-starlink-satellites.html.
33. "Starlink." Starlink, July 28, 2023. https://www.starlink.com/.
34. Andrews, Robin G. "Starlink Offers Internet Access in Times of Crisis, but Is It Just a PR Stunt?" *Scientific American*, March 18, 2022. https://www.scientificamerican.com/article/starlink-offers-internet-access-in-times-of-crisis-but-is-it-just-a-pr-stunt/.

35. Sokol, Joshua. "The Fault in Our Stars." *Science,* October 7, 2021. https://www.science.org/content/article/satellite-swarms-are-threatening-night-sky-creating-new-zone-environmental-conflict.
36. Dartnell, Lewis. "How Big a Problem Are Starlink Satellites for Astronomers?" *BBC Sky at Night Magazine,* November 10, 2022. https://www.skyatnightmagazine.com/space-science/spacex-starlink-problem-astronomy/.
37. Andrews, Robin G. "Satellites and Junk Are Littering Space and Ruining Our Night Skies." *New Scientist,* October 27, 2021. https://www.newscientist.com/article/mg25133581-700-satellites-and-junk-are-littering-space-and-ruining-our-night-skies/.
38. Boyle, Rebecca. "Satellite Constellations Are an Existential Threat for Astronomy." *Scientific American,* November 7, 2022. https://www.scientificamerican.com/article/satellite-constellations-are-an-existential-threat-for-astronomy/.
39. Mallama, Anthony, and Jay Respler. "Visual Brightness Characteristics of Starlink Generation 1 Satellites." ArXiv, 2022. Accessed June 21, 2023. arxiv.org/pdf/2210.17268.pdf.
40. "Enormous ('Mega') Satellite Constellations." Jonathan's Space Report, July 1, 2023. https://planet4589.org/space/con/conlist.html.
41. Howell, "Starlink Satellites."
42. "Vera C. Rubin Observatory—Impact of Satellite Constellations." Vera C. Rubin Observatory, August 25, 2022. https://www.lsst.org/content/lsst-statement-regarding-increased-deployment-satellite-constellations.
43. Hu, Jinghan A., et al. "Satellite Constellation Avoidance with the Rubin Observatory Legacy Survey of Space and Time." ArXiv, 2022. Accessed June 21, 2023. https://arxiv.org/abs/2211.15908.
44. Clery, Daniel. "Fallen Giant." *Science,* January 14, 2021. https://www.science.org/content/article/how-famed-arecibo-telescope-fell-and-how-it-might-rise-again.
45. Andrews, Robin G. "Arecibo's Collapse Sends Dire Warning to Other Aging Observatories." *Scientific American,* December 11, 2020. https://www.scientificamerican.com/article/arecibos-collapse-sends-dire-warning-to-other-aging-observatories/.
46. Cheatham, Amelia, and Diana Roy. "Puerto Rico: A U.S. Territory in Crisis." Council on Foreign Relations, September 29, 2022. https://www.cfr.org/backgrounder/puerto-rico-us-territory-crisis.
47. Varela, Julio R. "Puerto Rico Is Being Treated Like a Colony after Hurricane Maria." *The Washington Post,* September 26, 2017.
48. Marcos, Coral M. "Statehood or Independence? Puerto Rico's Status at Forefront of Political Debate." *The Guardian,* July 20, 2022.
49. Leonard, Daniel. "Hopes Fade for Resurrecting Puerto Rico's Famous Arecibo Telescope." *Scientific American,* October 20, 2022. https://www.scientificamerican

.com/article/hopes-fade-for-resurrecting-puerto-ricos-famous-arecibo-telescope/.

50. "About NSF." U.S. National Science Foundation (NSF). https://new.nsf.gov/about.
51. Billings, Lee. "This Report Could Make or Break the Next 30 Years of U.S. Astronomy." *Scientific American*, August 18, 2021. https://www.scientificamerican.com/article/this-report-could-make-or-break-the-next-30-years-of-u-s-astronomy/.
52. Witze, Alexandra. "Next Stop, Uranus? Icy Planet Tops Priority List for Next Big NASA Mission." *Scientific American*, April 19, 2022. https://www.scientificamerican.com/article/next-stop-uranus-icy-planet-tops-priority-list-for-next-big-nasa-mission/.
53. "Planetary Science and Astrobiology Decadal Survey 2023-2032." National Academies, April 19, 2022. https://www.nationalacademies.org/our-work/planetary-science-and-astrobiology-decadal-survey-2023-2032.
54. "Don't Look Up | Ending Scene." Nerd Clips HD, YouTube Video, December 25, 2021. https://www.youtube.com/watch?v=4-zv5Cvg6pM.

IX. The Cotton Candy Killers

1. Kelley, Jay W. "SPACECAST 2020, Volume 1." Defense Technical Information Center, (1992). Accessed June 20, 2023.
2. Mann, Adam. "How to Deflect Killer Asteroids with Spray Paint." *Wired*, February 7, 2013. https://www.wired.com/2013/02/painting-asteroids/.
3. "How Paintballs Could Save Earth from Giant Asteroid Impact." Space.com, October 26, 2012. https://www.space.com/18248-paintballs-asteroid-impact-deflection-video.html.
4. "Ion Propulsion." NASA, January 11, 2016. www.nasa.gov/wp-content/uploads/2015/08/ionpropfact_sheet_ps-01628.pdf.
5. Mizokami, Kyle. "The U.S. Navy's Railgun Is Nearly Dead in the Water." *Popular Mechanics*, April 27, 2020. https://www.popularmechanics.com/military/navy-ships/a32291935/navy-railgun-failure/.
6. "The Nobel Prize." Berkeley Lab, January 20, 2023. https://www.lbl.gov/people/excellence/nobelists/.
7. Lubin, Philip, and Alexander Cohen. "Planetary Defense Is Good—But Is Planetary Offense Better?" *Scientific American*, October 13, 2021. https://www.scientificamerican.com/article/planetary-defense-is-good-but-is-planetary-offense-better/.
8. Bruzek, Alison. "Teaching at the End of the World." *MIT Technology Review*, April 23, 2013. https://www.technologyreview.com/2013/04/23/178904/teaching-at-the-end-of-the-world/.
9. Bruzek, "End of the World."

10. Bruzek, "End of the World."
11. "Meteor Movie Review." *The Tech*, October 30, 1979. https://web.archive.org/web/20220418194111/http://tech.mit.edu/archives/VOL_099/TECH_V099_S0470_P003.pdf.
12. "Ames Technology Capabilities and Facilities." NASA, March 29, 2008. https://www.nasa.gov/ames-capabilities-facilities/.
13. "Ames Vertical Gun Range." NASA, 2010. https://www.nasa.gov/sites/default/files/files/ames-vertical-gun-range-v2010(1).pdf.
14. Mann, Adam. "NASA Brings Out the Big Gun for Asteroid Impact Science." *Wired*, August 19, 2013. https://www.wired.com/2013/08/a-scientist-and-his-gun/.
15. Mann, "NASA Brings Out the Big Gun."
16. Bennett, Jay. "NASA's High-Speed Cannon Used to Explore the Origins of Water on Earth." *Popular Mechanics*, April 26, 2018. https://www.popularmechanics.com/space/a20073974/nasas-high-speed-cannon-origins-of-water-on-earth/.
17. "NASA Ames Looks Back On Contributions to Apollo 11 Moon Mission." *CBS News*, July 16, 2019. https://www.cbsnews.com/sanfrancisco/news/nasa-ames-looks-back-on-contributions-to-apollo-11-moon-mission/.
18. Schultz, Peter H., and Stephanie N. Quintana. "Impact-Generated Winds on Mars." *Icarus*, 2017. Accessed June 21, 2023. https://www.sciencedirect.com/science/article/abs/pii/S0019103516305942.
19. Kornei, Katherine. "Deflecting an Asteroid Before It Hits Earth May Take Multiple Bumps." *The New York Times*, August 25, 2021.
20. Flynn, George, et al. "Momentum Transfer in Hypervelocity Cratering of Meteorites and Meteorite Analogs: Implications for Orbital Evolution and Kinetic Impact Deflection of Asteroids." *International Journal of Impact Engineering*, 2020. Accessed June 21, 2023. https://www.sciencedirect.com/science/article/abs/pii/S0734743X1930661X.
21. Mann, "Big Gun."
22. "DART's Small Satellite Companion Takes Flight Ahead of Impact." JHUAPL. September 15, 2022. https://www.jhuapl.edu/PressRelease/220915-dart-cubesat-companion-liciacube-deploys-ahead-of-impact.
23. "NASA Announces Pending Departure of Science Associate Administrator." NASA, September 13, 2022. https://www.nasa.gov/press-release/nasa-announces-pending-departure-of-science-associate-administrator/.
24. Chang, Kenneth. "At NASA, Dr. Z Was OK With Some Missions Failing." *The New York Times*, January 12, 2023.
25. "We Choose to Go to the Moon . . ." Wikimedia Foundation. Last modified May 29, 2023. https://en.wikipedia.org/wiki/We_choose_to_go_to_the_Moon.

X. Ashes to Ashes

1. "NASA Confirms DART Mission Impact Changed Asteroid's Motion in Space." NASA, October 11, 2022. https://www.nasa.gov/press-release/nasa-confirms-dart-mission-impact-changed-asteroid-s-motion-in-space.
2. Keeter, Bill. "DART Sets Sights on Asteroid Target." NASA, September 7, 2022. https://www.nasa.gov/directorates/smd/dart-sets-sights-on-asteroid-target/.
3. Rivkin, Andy. Twitter, September 6, 2022. https://twitter.com/asrivkin/status/1567143925743230976?s=21&t=uDMd7_jW02f-XCFpoPxKaA.
4. Rivkin, Andy. Twitter, September 26, 2022. https://twitter.com/asrivkin/status/1574175414389395457.
5. Chabot, Nancy. Twitter, September 26, 2022. https://twitter.com/nlchabot/status/1574220635441876992.
6. "Planetary Names." USGS. https://planetarynames.wr.usgs.gov/Page/Categories.
7. "Uranus Moons." NASA, September 15, 2022. https://solarsystem.nasa.gov/moons/uranus-moons/in-depth/.
8. "Dhol Saxum." USGS, January 26, 2023. https://planetarynames.wr.usgs.gov/Feature/16161.
9. Dobrijevic, Daisy, and Elizabeth Howell. "Comet Hale-Bopp: Facts about the Bright and Tragic Comet." Space.com, April 18, 2022. https://www.space.com/19931-hale-bopp.html.

XI. Dust to Dust

1. "SAOO." Twitter, September 27, 2022. https://twitter.com/SAAO/status/1574688994201255936?s=20&t=qzBvmgF7rnLbFwBAhEFD5Q.
2. "Atlas Project." Twitter, September 27, 2022. https://twitter.com/fallingstarifa/status/1574583529731670021?s=46&t=ZajqfNk7kxmzZa-cz_VTfQ.
3. Project, Atlas. "SOAR Telescope Catches Dimorphos's Expanding Comet-Like Tail After DART Impact." NOIR Lab, October 3, 2022. https://noirlab.edu/public/news/noirlab2223/.
4. Bardan, Roxana. "NASA Confirms DART Mission Impact Changed Asteroid's Motion in Space." NASA. October 11, 2022. https://www.nasa.gov/press-release/nasa-confirms-dart-mission-impact-changed-asteroid-s-motion-in-space/.
5. Dyar, Darby, Bruce Betts, and Sarah Al-Ahmed. "The Case for Saving VERITAS." The Planetary Society, May 10, 2023. https://www.planetary.org/planetary-radio/2023-the-case-for-saving-veritas.
6. Rabie, Passant. "NASA Budget Request Is a 'Soft Cancellation' of Venus Mission, Experts Say." *Gizmodo*, March 23, 2023. https://gizmodo.com/nasa-budget-soft-cancellation-veritas-venus-mission-1850245199.

7. Drake, Nadia. "Artemis I Launches U.S.'s Long-Awaited Return to the Moon." *Scientific American,* November 16, 2022. https://www.scientificamerican.com/article/artemis-i-launches-u-s-s-long-awaited-return-to-the-moon/.
8. Sheetz, Michael. "Investing in Space: Where Does NASA Go from Here?" CNBC, September 29, 2022. https://www.cnbc.com/2022/09/29/investing-in-space-where-does-nasa-go-from-here.html.
9. Drake, "Artemis I Launches."
10. Jones, Andrew. "China to Target Near-Earth Object 2020 PN1 for Asteroid Deflection Mission." *SpaceNews,* July 12, 2022. https://spacenews.com/china-to-target-near-earth-object-2020-pn1-for-asteroid-deflection-mission/.
11. Steigerwald, William. "Early Results from NASA'S DART Mission." NASA. December 15, 2022. https://www.nasa.gov/feature/early-results-from-nasa-s-dart-mission/.
12. Kareta, Theodore, et al. "Ejecta Evolution Following a Planned Impact into an Asteroid: The First Five Weeks." ArXiv, (2023). Accessed November 29, 2023. arxiv.org/abs/2310.12089.
13. Andrews, Robin. "Remember When NASA Crashed into an Asteroid? It Had Some Unintended Consequences." *National Geographic,* August 10, 2023. https://www.nationalgeographic.com/premium/article/nasa-dart-mission-created-unintended-consequences-swarm-of-boulders?loggedin=true&rnd=1701275214516.
14. "Dragonfly." JHUAPL. https://www.jhuapl.edu/work/projects/dragonfly.

Epilogue: What Do We Say to the God of Death?

1. "Nacreous Clouds." Met Office. https://www.metoffice.gov.uk/weather/learn-about/weather/types-of-weather/clouds/other-clouds/nacreous.

INDEX

Page numbers in *italics* refer to illustrations.